AI AND UX

WHY ARTIFICIAL INTELLIGENCE NEEDS USER EXPERIENCE

Gavin Lew
Robert M. Schumacher Jr.

Apress®

AI and UX: Why Artificial Intelligence Needs User Experience

Gavin Lew
S Barrington, IL, USA

Robert M. Schumacher Jr.
Wheaton, IL, USA

ISBN-13 (pbk): 978-1-4842-5774-6
https://doi.org/10.1007/978-1-4842-5775-3

ISBN-13 (electronic): 978-1-4842-5775-3

Managing Director, Apress Media LLC: Welmoed Spahr
Acquisitions Editor: Shiva Ramachandran
Development Editor: Rita Fernando
Coordinating Editor: Nancy Chen

Cover designed by eStudioCalamar

Distributed to the book trade worldwide by Springer Science+Business Media New York, 1 New York Plaza, New York, NY 100043. Phone 1-800-SPRINGER, fax (201) 348-4505, e-mail orders-ny@springer-sbm.com, or visit www.springeronline.com. Apress Media, LLC is a California LLC and the sole member (owner) is Springer Science + Business Media Finance Inc (SSBM Finance Inc). SSBM Finance Inc is a **Delaware** corporation.

For information on translations, please e-mail booktranslations@springernature.com; for reprint, paperback, or audio rights, please e-mail bookpermissions@springernature.com.

Apress titles may be purchased in bulk for academic, corporate, or promotional use. eBook versions and licenses are also available for most titles. For more information, reference our Print and eBook Bulk Sales web page at http://www.apress.com/bulk-sales.

Any source code or other supplementary material referenced by the author in this book is available to readers on GitHub via the book's product page, located at www.apress.com/9781484257746. For more detailed information, please visit http://www.apress.com/source-code.

Printed on acid-free paper

Contents

About the Authors

Gavin Lew has over 25 years of experience in the corporate and academic environment. He founded User Centric and grew the company to be the largest private UX consultancy in the United States. After selling the company, he continued to lead a North American UX team to become one of the most profitable business units of the parent organization. He is a frequent presenter at national and international conferences and the inventor of several patents. He is an adjunct professor at DePaul and Northwestern universities. Gavin has a Masters in Experimental Psychology from Loyola University and is currently the Managing Partner of Bold Insight, part of ReSight Global, a globally funded UX consulting practice across North America, Europe, and Asia.

Robert M. Schumacher Jr. has more than 30 years of experience in academic, agency, and corporate worlds. He co-owned User Centric with Gavin from its early stages until it was sold to GfK in 2012. While at User Centric, Bob helped found the User Experience Alliance, a global alliance of UX agencies. Also, he founded User Experience Ltd, a UX agency in Beijing. He is co-founder, co-owner, and Managing Partner of Bold Insight, part of ReSight Global, a global UX company. Bob was the editor of and contributor to *The Handbook of Global User Research* (2009). He has several patents and dozens of technical publications, including user interface standards for health records for the US government. He also is an Adjunct Professor at Northwestern University. Bob has a Ph.D. in Cognitive and Experimental Psychology from the University of Illinois at Urbana-Champaign.

Preface

Our perspectives and biases

We have both been around long enough to see technology grow from mail order Heathkit computers sold from the ads in *Popular Mechanics* to the exponential and ubiquitous presence that technology has in our lives. And we're not that old.

Because computing advances came at us so fast, the user was often seen simply as an input/output device. The user had to adapt to the system rather than building the system around the user's skills, knowledge, and capabilities. What has driven us professionally and personally is that what we do as user experience (UX) professionals matters in the lives of people. While working together at Ameritech (a "Baby Bell" regional phone company) in the 1990s, we were involved with making products more successful by focusing on the UX. We evaluated products and often paused to shake our heads and think, "Why would anyone design the product *this way?*"

To put it simply, *we believe experiences matter*. We want to make the world a little easier for people.

Our perspective on how AI can be more successful is admittedly and unashamedly from a UX point of view. AI needs a focus on UX to be successful.

UX has a history with DNA strands from several places, most notably psychology. We were both trained as experimental psychologists, but along the way we had some significant exposure to computer science and AI in particular. It was easy to be seduced by programs like Eliza that seemingly converses with you or believe in proclamations of the glorious future that AI would bring because the culture had such limited experience in computing technology. It was mysterious and magical. But as the scales fell away, we saw it for what it was: *code*. What we thought was smarter computing was simply clever code that often fooled the user. That's not to say that computer scientists weren't sincere—they understood that these were not true thinking machines. But those who did not understand (journalists and the rest of us) may have gotten over our skis as to what AI could do. The payment for overhyping in the eyes of the public was to lose faith and trust in AI.

Part of that loss of trust, as shown in several examples, was due to the fact that AI was often unpolished. The thinking was often this: the AI engine works, yay! But there was not a lot of attention to how the end users benefitted from the AI tool. Humans are impatient and fickle creatures; unless they are going to see the benefit very early on, they often will not invest the time or attention needed to appreciate the AI brilliance. And this is what happened. A bad experience with AI poisons the well. People won't go back. What's worse is that those users will often paint a whole class of AI-enabled products with the same brush.

These failed experiences in AI bore a remarkable similarity to failures we saw due to poor UX in product design. Bad experiences meant poor perceptions, lack of usage, and ultimately declining success.

But with AI, we often witnessed a gap where those who normally have strong opinions would give AI technology the benefit of the doubt, as it was beyond their area of expertise. If AI is to be successful, the design matters. The UX matters. How people would interact with AI matters. We believe UX *can* help; that's the main point of the book!

About this book

AI and UX are expansive, and we are unable to plumb the depths of either of them. We try to stay close to what we know and what we thought was relevant to make our points.

We don't wish to lump all AI applications together. In this book, we mainly center on AI that directly touches people doing tasks—whether it's at home, in the office, or on the go. The focus is not on financial trading algorithms or epidemiological modeling or the AI that runs in the background of industrial automation that does not rely on or present information to people. Our focus in this book is on the AI that most of us will experience—specifically the AI that is experienced by us all in commonly used applications.

We employ dialogs in the book that allows us to be more casual and communicate more as you would talking to a friend or colleague. Sometimes we use the dialogs to make the point, other times to reinforce it. It is our hope that this technique is successful in highlighting our key points.

The book is laid out so that in the early chapters, we describe the relevant history of both AI and UX—and how that history intertwined in the lives of some very influential researchers. We then lay out the specific problem in Chapter 4. Chapter 5 is where we are prescriptive about how UX can benefit AI through the user-centered design model.

Acknowledgments

Any endeavor of this sort does not happen without support from an awesome collection of people. What started as a number of discussions with a colleague, Dan Delaney, over lunch perhaps 5 years ago formed the thesis that drove the book, which was to not focus on the algorithms but the impact AI could have on everyday people. But we knew that this could only occur if AI was successful. We spent our careers shaping the design of products to better fit those who would use and benefit. *What was the applicability to AI?* Those lunchtime discussions became a concept used in the book where Bob and Gavin have a dialog. We wanted to use these discussions to bridge the gap between history and opinion to give the reader something tangible and, hopefully, an insightful perspective. Dan is acknowledged as a major influence, and we wish he was able to have played a larger role.

We also would like to thank Ethan Lew, Gavin Lew's eldest son who is studying computer science. When Gavin gave a presentation on some of the key messages in the book, Ethan called the main points *rudimentary*. Somehow Gavin was able to recover when one of the founding fathers of cognitive science and subsequently UX, Don Norman, told Gavin after listening to the presentation, "Rudimentary means it has appeal to a wider audience. Write the book." Both Ethan and Don helped energize our belief that the time was right for the discipline of UX and the development of AI to come together and design better outcomes. We can all play a role in shaping AI to be more successful.

We also want to express our sincere gratitude to Claudette Lew, Gavin's spouse, for countless hours reading revisions for clarity. We are also indebted to our team at Bold Insight. Their perspectives were invaluable in shaping our ideas and crafting our book. Their insight during brainstorming meetings to review for flow and comprehension helped us bring this whole thing together.

We also want to acknowledge the early support from JD Lavaccare who helped craft our early drafts. Our editorial team at Apress was ever-present and always helpful—their ideas sharpened our language and clarified our thinking. Lastly, many thanks to our families who put up with us while we were catatonic facing a blank page. We did our best to stay with this moving target. Any errors or omissions are entirely our own.

Introduction to AI and UX

There and back again

Name any field that's full of complex, intractable problems and that has gobs of data, and you'll find a field that is actively looking to incorporate artificial intelligence (AI). There are direct consumer applications of AI, from virtual assistants like Alexa and Siri to the algorithms powering Facebook and Twitter's timelines, to the recommendations that shape our media consumption habits on Netflix and Spotify. MIT is investing over a billion dollars to reshape its academic program to "create a new college that combines AI, machine learning, and data science with other academic disciplines." The college started September 2019 and will expand into an entirely new space in 2022.[1] Even in areas where you'd not expect to find a whiff of AI, it emerges: in the advertising campaign to its new fragrance called Y, Yves Saint Laurent

---

[1]Knight, Will (2018). "MIT has just announced a $1 billion plan to build a new college for AI." MIT Technology Review. Last updated October 15, 2018. Last accessed June 2, 2020. www.technologyreview.com/f/612293/mit-has-just-announced-a-1-billion-plan-to-create-a-new-college-for-ai/.

© Gavin Lew, Robert M. Schumacher 2020
G. Lew and R. M. Schumacher, AI and UX,
https://doi.org/10.1007/978-1-4842-5775-3_1

showcased a model who is a Stanford University graduate and a researcher in machine vision.[2] The commercial showcases AI as hip and cool—even displaying lines of Python code, as well as striking good looks to sell a fragrance line. AI has truly achieved mainstream appeal in a manner not seen before. AI is no longer associated with geeks and nerds. AI now sells product.

THE INCREDIBLE JOURNEY OF AI

GAVIN: *The Hobbit, or There and Back Again* by J. R. R. Tolkien tells of Bilbo Baggins' incredible journey and how he brought his experience back home to tell his tale. That novel opened the door to science fiction and fantasy for me.

BOB: Same for me as well. As I got older, science fiction became more real and approachable. What was once fantasy is now tangible. Consider artificial intelligence. It has gone farther and faster than I would have believed even a decade ago. And while AI did not encounter dragons, wizards, and elves as in *The Hobbit*, AI did have perils and pitfalls on the journey.

GAVIN: Like Bilbo, the story of AI is a journey that carries lessons to be learned. I think Tolkien's story was not about where the future can take you, but to not forget what the past can teach, inform, and make better. *The Hobbit* was the prelude to an even bigger story that became *The Lord of the Rings*. AI may indeed have a great future, but getting it right will require some new thinking; this book is a UX researcher's tale on AI.

The point　AI has a long history. Learning from its mistakes made in the past can set the AI of today for success in the future.

The world, both inside and outside the tech industry, is abuzz with AI.

There must be more to AI than being a company's newest cool thing and giving fodder to marketers. The foundation that unlocks the massive opportunity to answer questions and make human lives easier is the power and intrigue of AI. But its potential is dependent upon having technology work. Because when technology does not work, there are consequences.

---

[2]James, Vincent (2017). "AI is so hot right now researchers are posing for Yves Saint Laurent." The VERGE. Last updated August 31, 2017. Last accessed August 12, 2019. www.theverge.com/tldr/2017/8/31/16234342/ai-so-hot-right-now-ysl-alexandre-robicquet.

OVERHYPED FAILURES HAVE CONSEQUENCES

GAVIN: The excitement around AI is white-hot. As an example, in health care, Virginia Rometty, former CEO of IBM, said that AI could usher in a medical "Golden Age."[3] AI is in the news everywhere.

BOB: When one thinks of overhyped environments, I think of another Golden Age: the tulip craze in 17th-century Holland. Investing in tulip bulbs became highly fashionable, sending the market straight up. As the hype grew, a speculative bubble emerged where a single bulb hit 10 times an average worker's annual salary.[4] Inevitably the market failed to sustain the crazy prices and the bubble burst.

GAVIN: Like "tulip mania," the hype around AI is high, if not "irrationally exuberant." But what may be surprising to many is that this is not the first time AI has been hyped up. The boom years of AI in the late1950s came to a crash in the decade that followed. Virtually all funding for AI was cut and it took another couple of decades to see investment resume.

BOB: This slash of funding all things AI spanned a decade. As research began to move into robotics, the accompanying hype around robots led to another crash in the 1980s. AI's history is long and has seen peaks and valleys. My hope is that today's exuberance we will remember the lessons from the past, so this new era of AI will see a more successful future.

The point   Failures have implications and have occurred more than once with AI. Learning from mistakes made in the past can set the AI of today for success in the future.

Artificial intelligence

The term "artificial intelligence" was originated by computer scientists John McCarthy, Marvin Minsky, Nathaniel Rochester, and Claude Shannon in 1955. They defined AI as "…making a machine behave in ways that would be called intelligent if a human were so behaving."[5] Of course, this still leaves the

---

[3]Strickland, Eliza (2019). "How IBM Watson Overpromised and Underdelivered on AI Health Care." IEEE Spectrum. Last updated April 2, 2019. Last accessed June 2, 2020. https://spectrum.ieee.org/biomedical/diagnostics/how-ibm-watson-overpromised-and-underdelivered-on-ai-health-care.

[4]Goldgar, Anne (2008). Tulipmania: money, honor, and knowledge in the Dutch Golden Age. University of Chicago Press.

[5]Press, Gil. "Artificial Intelligence (AI) Defined." Forbes. Published August 27, 2017. Accessed June 2, 2020. www.forbes.com/sites/gilpress/2017/08/27/artificial-intelligence-ai-defined/#44ac4d9f7661.

definition of AI widely open to interpretation based on the subjective definition of what is "intelligent" behavior. (Needless to say, we know a lot of humans who we don't think behave intelligently). AI's definition remains elusive and changeable.

The question "What is intelligence?" is outside the scope of this book, fraught as it is with philosophical complications. But, in general, we would support a version of the original definition of artificial intelligence. The domains contained within artificial intelligence all share a common thread of automating tasks that might otherwise require humans to exercise their intelligence.

There are alternative definitions, such as the one offered by computer scientist Roger Schank. In 1991, Schank laid out four possible definitions for AI as:

1. Technology that can divine insights with no direction from humans;

2. "Inference engines" that can be fed information about any particular field and calculate proper courses of action;

3. Any technology that does something that has never been done by technology before; and

4. Any machine capable of learning.

We see these as four different ways of defining "intelligence." Schank endorses the fourth definition, thereby endorsing the idea that learning is a necessary part of intelligence.

For the purposes of this book, we will not be using Schank's definition of AI—or anyone else's. Doing so would require us to redefine past AI systems, and even some present AI systems, as outside the realm of AI, which we do not intend to do. Machine learning is often a central part of AI, but it's fairly rare. Plenty of AI systems aren't great at learning on their own, but they can still accomplish tasks that many would consider intelligent. In this book, we want to discuss as many applications of AI as possible, whether they are capable of learning or not. So, we will define **artificial intelligence** in the most broad way possible.

Definition **Artificial intelligence**, or AI, has a meaning that is much contested. For our purposes, artificial intelligence is any technology that appears to adapt its knowledge or learns from experiences in a way that would be considered intelligent.

A wide variety of technologies can be considered part of AI. So, we've adopted a broad definition of the term. Today's AI applications can do many tasks that were previously thinkable or would require exceptional effort. Machine translators, like Google Translate, can translate between hundreds of languages in a split second at adequate quality for many applications. Medical and business AI can analyze large swaths of data and output insights that can help professionals do their jobs more efficiently. And, of course, virtual assistants allow users to complete tasks such as sending messages and ordering products with that most natural of interfaces—voice.

The emergence of artificial intelligence is closely timed with that of user experience, both coming at the advent of the computer age. We'll go over it in more detail in Chapter 2 but suffice it to say that some of the most important innovations in AI and computing in general—neural networks, Internet gateways, graphical user interface (GUI), and more—were made possible by the work of psychologists-turned-computer scientists. UX is heavily influenced by psychology, and many of the psychologists' questions were focused on something akin to human-computer interaction (even if some of their work predates the advent of that particular field).

The point AI's definition has centered on using computational methods to accomplish intelligent tasks. We won't concern ourselves with which complex tasks are "intelligent" and which aren't. We're more concerned with trying to help make AI more successful by applying a UX-centered approach.

User experience

Don Norman coined the term "user experience" in 1993, while working for Apple. In a video for his research organization, the Nielsen Norman Group, Norman describes user experience as a holistic concept, incorporating the entirety of an experience of buying and using a product.[6] He presents the example of purchasing a computer in the 1990s, imagining the difficulty of lugging the computer box into one's car and the intractability of the computer's setup process. He implies that these experiences—even as they are seemingly divorced from the actual functionality of the device—can affect the user's overall perception of the device's functionality. This reveals the all-encompassing nature of **user experience**.

---

[6]NNgroup. "Don Norman: The term 'UX.'" YouTube video, 01:49. Posted July 2, 2016. Accessed June 2, 2020. www.youtube.com/watch?v=9BdtGjoIN4E.

Definition **User experience**, or UX, asks designers to look at new technologies as experiences, not products. UX designers use models based on social sciences, especially psychology, to design experiences so that users can effectively and efficiently interact with things in their world.

UX vis-à-vis AI

If technology doesn't work for people…it doesn't work.[7]

This was an old marketing slogan used by Ameritech, which was a Regional Bell Operating Company (RBOC). Ameritech formed after the breakup of the Bell System monopoly in 1984 where AT&T provided long-distance telephone service while the other RBOCs provided local telephone service. Many know RBOCs as Pacific Bell (Pac Bell), SBC (Southwestern Bell), NYNEX, BellSouth, US West, and so on. The Ameritech slogan represented the work of a small team of 20 human factors engineers and researchers managed by Arnie Lund. Arnie was a mentor to us (the authors) and to dozens who learned under his leadership; we saw the evolution of human factors to user experience while working for Arnie.

The role of this team was to make products "work" for the user. It seems rather simple that products, of course, need to work as they were intended. But the key is not whether they work for an engineer, but for the user— someone who bought the product or possibly received it as a gift. Think about some products you purchased. Think about the ones with batteries or those that plug into the wall or even connect to the Internet. Would you say the setup experience was easy? Unfortunately, there are a lot of products that just make us shake our heads and ask who made it so hard to use? The Ameritech slogan encapsulated the research and design that was needed not simply to integrate technology into new products but to transform experiences for people as a critical criterion of success. Incorporating a user-centered approach was not the norm in 1995. UX was not "table stakes" as it is now, but it was such a unique selling point for Ameritech that they featured it in broadcast TV ads.

---

[7]Oddly, Ameritech never filed for a trademark for this slogan. Ultimately, a registered trademark was given to User Centric, Inc., a company that was owned by both authors, in 2005.

IF AI DOESN'T WORK FOR PEOPLE, IT DOESN'T WORK

GAVIN: The human factors team at Ameritech, under Arnie Lund, was an amazing group at a special time. For me, it was my first experience applying psychology, research, and user-centered design to make positive changes in a product.

BOB: The team included some spectacular minds who were allowed the freedom to make products useful, usable, and engaging at a time before Apple grabbed hold of the *Think different* mantra.

GAVIN: I remember Ameritech as an organization that had tens of thousands of employees, but yet an entire TV advertising campaign was based on the work of a very small team.

BOB: The commercials were referred to as the Ameritech Test Town ads.[8] They illustrated how Ameritech's human factors team would test new technology with people in everyday places like diners, cars, and barber shops to ensure that products did not just function, but people could actually use them.

GAVIN: I remember the ads were memorable and scored high on traditional audience measures. Everyone in the Midwest in the mid-1990s knew the ads. As I remember, the campaign had an ad recall measure that was off the charts good. But, when only half of the respondents thought the ad was about Ameritech; half mistakenly said it was AT&T.

BOB: That's true. I'm sure the marketing people were chagrined, but to us, it didn't matter. The point was not to show off some amazing new technology; it was not about slick cutting-edge features. The real message was about the *experience*. If the person who bought the device could not use it, then nothing mattered. My 88-year-old father still contacts me every other day about how frustrated he is with his computer.

GAVIN: This was the essence of what made a good UX. And honestly, we have still not come all that far today. Sure, the technology has accelerated, but if the person can't book the trip on a website or program their digital watch easily, then the device is just a digital pet rock because it will get little to no use. Around 20 years have gone by and while there is certainly an increased awareness of UX, our lives are still as—or perhaps even more—frustrated by products and services in our world.

---

[8]See www.youtube.com/watch?v=1KUZPR52uCU for one of the commercials. Last accessed June 2, 2020.

BOB: As devices get "smarter," some people might think they will reduce our frustrations. In fact, there is a school of thought that user interfaces will just fade into the background. They will be the underlying technology, increasingly invisible because they are so intuitive. I am not entirely sold on this yet, but we'll come back to that. For now, as we read on about AI and all that AI promises, *the emphasis on how people experience AI seems to be missing.*

The point   Product developers increasingly recognize that good design matters. Understanding how people interact is critical to product success.

The Ameritech "Test Town" commercials showed short vignettes of futuristic technology in everyday life. In one, there was a coffee shop where patrons wore devices; one of those devices frustrated the user because it kept flashing 12:00. (Something many of us can relate to!) The premise was that there were people who worked at Ameritech who were making products easier to use; products that did not just work, but "worked for people."

These ads brought attention to the work being done behind the scenes to improve the utility and usability of the technology in Ameritech's new products and services. The field of UX is focused on understanding and improving the connection between humans and technology so that the experience could be more than satisfying.

When you think of a user experience, consider adjectives and adverbs that describe the interaction, such as in Figure 1-1. When you design a product or service, you don't want it to simply be satisfying; sometimes satisfying just means 'good enough.' But isn't that what you hear about? Customer *satisfaction?* But we argue that a product's success tends to need more than a satisfying experience. When one focuses on the user experience of a product, the interaction design needs to do much more. When you think of something that you really enjoy using, the words you might use are that is *addictive, fun, engaging, intuitive,* and so on. These are the descriptors that make a user experience great. For success, we must strive for more than satisfaction. We need product to be associated with UX adjectives and adverbs.

Figure 1-1. Adjectives and adverbs that are used to describe interactions. These UX adjectives and adverbs raise the bar for an experience of a product beyond "satisfying"

We, as UX professionals, can't help but look at AI as simply an application. It is not just about the promise of what AI can do for us. We believe that peering at AI through a UX prism will be instructive; a good UX experience is vital to ensure success and future propagation of AI. UX is our area of expertise, and the thesis of this book is to propose applying UX principles to the design and development of user interfaces to artificial intelligence applications.

In the past, AI has been designed by thinking about the functions and the code: *What if we could get AI to do X?* Designers and developers have big dreams for a mind-bending set of AI applications and set out to achieve them. What we think they ignore all too often is this: *Once AI can do X, what will it be like to use AI to do X?* In other words, developers need to think about what the *experience* of using their AI product will be like—even in the early stages, when that product is just a big idea. It's great to dream that the bot will convert speech to text to action, but if the speech is in a crowded bar and the speaker just had dental work, how useful is the bot? At this point, we posit that an essential element in AI's success hinges on understanding and improving the *user experience*—not on giving AI all manner of new functions. Most AI applications already have plenty of useful functions, but what good are functions if the user can't use them or doesn't know how to access them?

> # HAVING A GOOD INITIAL EXPERIENCE
> # GOES A LONG WAY

BOB: So much time and effort goes into product design. When a bad experience happens, sometimes I think how close the designers came to getting it right. What could the team that made it have done to get it right? There is often a fine line between success and failure.

GAVIN: Yeah, think of voice-enabled calling in a car. The auto manufacturers have had this around for a decade. But now, almost everyone has a phone; how many use the voice feature in their vehicles? The old adage of "fool me once, shame on you. Fool me twice, shame on me" might apply to human to human interactions, but human to AI interactions is more like, "Fail me once and you tend to not try again."

BOB: Exactly, imagine a mom driving a car full of kids to a soccer game. She tries to use voice calling in the car. If the mom hears, "I don't understand that command," do you think she will ever try again? Will she realize the ambient background noise (i.e., children playing) might have interfered with AI's ability to understand? Most won't.

GAVIN: Applying this logic to all the technology around us, this need for a good user experience goes beyond voice calling—think about the effort that goes into the design of the 500+ features in a BMW 540, for example. So much time and cost goes into building these features. But, how many do people actually use? Just because a feature's there doesn't mean it's useful or usable.

BOB: UX focuses on more than how the feature works. Half the battle is helping people get to the feature. Once accessed, does the feature map to how people expect it to work? These are core principles of good design. AI is not a panacea. Understanding how users will interact with this output is the experience. And that where focus on the UX is key.

The point   Products embedded with new technology do not automatically ensure success—a positive interaction is essential.

UX framework

With UX described as an important driver for success, how UX integrates into AI-enabled products starts with the introduction of a **UX framework** which will lay the foundation for topics in this book.

Definition   The **UX framework** is our method of considering user experience while designing an AI application. This framework is rooted in classic user-centered design where the user is at the center, not technology.

AI-UX principles

In order to understand how a UX framework can be applied to AI, we'll consider these three AI-UX principles: context, interaction, and trust. These are independent dimensions that make up our UX framework for AI. We will cover this model in depth in later chapters, but it is instructive to give a small tasting of these now. See Figure 1-2.

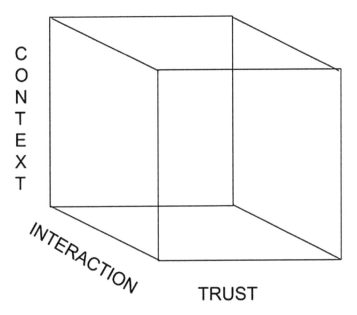

Figure 1-2. AI-UX principles that need to be addressed when designing AI products and services

---

The point The next wave of AI needs to be designed with a UX framework in mind or risk the further limiting of acceptance. Good UX for AI applications will propel growth.

Context

In recent years, IBM spent $4 billion on acquisitions and expenses in the healthcare sector, mostly to boost the capabilities of its new medical AI, Watson Health.[9] The results have been mixed. Watson Health has shown some incredible promise, but it's also stagnated. *The Wall Street Journal* published a scathing article about Watson Health's failures in 2018.[10] The article alleged that "more than a dozen" clients have cut back or altogether dropped their usage of Watson Health's oncology (cancer treatment) programs and that there is little to no recorded evidence of Watson Health's effectiveness as a tool for helping patients.

In 2017, Watson Health's ability to create cancer treatment plans was tested for agreement with doctors' recommended treatment plans in both India and South Korea. When Watson was tested on lung, colon, and rectal cancer patients in India, it achieved agreement rates ranging from 81% to 96%. But when it was tested on gastric cancer patients in South Korea, it achieved just 49% agreement. Researchers blamed the discrepancy on the diagnostic guidelines used by South Korean doctors, which differed from those Watson was trained on in the United States.[11]

---

[9]Agence France-Presse. "IBM Buys Truven Health Analytics for $2.6 Billion." IndustryWeek. Last updated February 16, 2016. Accessed June 2, 2020. www.industryweek.com/finance/ibm-buys-truven-health-analytics-26-billion.

[10]Hernandez, Daniela and Ted Greenwald. "IBM has a Watson dilemma." *The Wall Street Journal.* Last updated August 11, 2018. Accessed June 2, 2020. www.wsj.com/articles/ibm-bet-billions-that-watson-could-improve-cancer-treatment-it-hasnt-worked-1533961147.

[11]Ramsey, Lydia. "Here's how often IBM's Watson agrees with doctors on the best way to treat cancer." Business Insider India. Last updated June 2, 2017. Accessed June 2, 2020. www.businessinsider.in/Heres-how-often-IBMs-Watson-agrees-with-doctors-on-the-best-way-to-treat-cancer/articleshow/58965531.cms.

DON'T LIMIT AI TO IMITATING HUMAN BEHAVIORS

BOB: So Watson learned how to diagnose and recommend cancer treatments using a US dataset. Everyone cheered when Watson recommended treatment plans that US doctors recommended. But, when applied to South Korean cases, it missed the mark. But, is this the criterion for success? Replicating what US doctors do?

GAVIN: This is the point! AI should not merely replicate. The fact that AI found a difference is noteworthy. Perhaps we need to change our thinking. AI found a difference. And *this* is the insight. AI is raising its hand and effectively asking, "What are South Korean oncologists doing that US oncologists are not? And why are they making those decisions?" Instead, many interpreted this as an inability to imitate human decision making and therefore Watson failed.

BOB: Yes! Improving health outcomes is the goal—not whether a computer's recommendations correlate to a human. AI's contribution asks us to investigate the difference in treatments. This might advance care by looking at the differences to find factors that improve outcomes. It is the difference that might help.

The point   We limit AI by determining success on whether its outcomes correlate with human outcomes. This is merely the first step where AI can identify differences. Knowledge is furthered when this insight spurs more questions. This leads to the ultimate goal: better health outcomes. When we make the goal about replicating human outcomes, we do a disservice to AI.

Ideally, Watson should be applauded for identifying that there is a difference in treatment plans between US and South Korean cases. AI does not have to solve the entire problem. Finding a difference that was previously unknown is a big step. AI alerted us to a difference. Now, we can investigate and possibly save lives.

This is a good example of engaging more than just programmers with the challenge of making AI solve problems. Let's take control and design a product with AI as a team. Bring in product teams, programmers, oncologists, and even marketing to approach the problem and not assume AI will figure it out by itself.

We'll talk more about AI examples, like Watson Health in Chapter 3, but for now, suffice it to say that it is one example of how AI can stagnate without an awareness of **context**.

Definition   **Context** includes the outside information that AI can use to perform a task. It includes information about the user and why they are making the request, as well as information about the external world.

■ **The point**   Those who work on AI-enabled products need to understand the context of its output—what its output means, how it compares to objectives and expectations, and more.

Interaction

Gavin's college roommate at the University of California-San Diego was Craig Nies, who studied computer science. He went on to help develop Falcon, a pioneering AI algorithm that detected credit card fraud in the early 1990s. Gavin talked to Craig while working on this book, and Craig explained how he used a type of AI known as a neural network to detect suspicious purchases. Falcon incorporated variables like geographic location and types of stores to give each transaction a score. High-scored transactions meant a phone call to the cardholder and a possible canceled transaction.

This worked well in many situations, including situations of actual fraud. But it had trouble with false positives, in which it marked non-fraudulent purposes as fraudulent. One particularly problematic case was international trips. Back in the 1990s, not everyone had a cell phone, and even those cell phones that were around had difficulties with international calling. If you didn't have your working phone on you, or the credit card company neglected to call, you were liable to lose your credit card for your entire international trip. Of course, this could spell disaster.

Today, there is only one additional step involved. The credit card companies still use similar fraud detection mechanisms, identifying likely fraudulent purchases by geographic location and store type, among other factors. But now, thanks to the ubiquity of smartphones, not to mention the increasing coverage of mobile data and Wi-Fi networks, your credit card company can send an alert to your phone that asks whether or not you made the suspicious purchase in question. If you did, you can tap *Yes* and your purchase will go through just fine.

That additional **interaction** with the user makes a world of difference.

■ **Definition**   **Interaction** refers to AI engaging the user in a way in which they can respond. That engagement could come in many forms: a message in the AI's interface, a text message, a push notification to their smartphone, etc.

When the AI system makes its conclusion that the purchase is likely fraud, it doesn't act immediately to cancel the transaction and lock down the card. Instead, the AI algorithm has a convenient way to reach out to the user and make sure that the users don't object to it taking this action. While this request for user consent is not foolproof (e.g., perhaps the user's phone has been stolen alongside the card, or it is out of battery), it works better than the old method of giving the user a phone call. It's a more effective interaction—a necessity when the possible impacts are so great.

The point Before AI takes a potentially impactful action on a user's behalf, it should attempt to interact with the user. Communication is key. This is the experience that AI needs. The interaction needs to be designed.

Trust

When you think of **trust** relative to a device interaction, the initial expectation is that the device does what it is supposed to do. But from a UX perspective, **trust** can go further. Here is an example. If you are familiar with the iPhone, when you hear the Siri startup beep, what is your reaction? You might recoil a little: *Ugh, I accidentally pressed it again.* But why does Siri deliver such a negative visceral reaction in many of us? It's supposed to be a helpful tool, after all. Why don't we *trust* it?

Definition **Trust** is when users feel that an AI system will successfully perform the task that a user wants it to perform, without any unexpected outcomes. Unexpected outcomes can include performing additional (unnecessary or unhelpful) tasks that the user did not ask for or breaching the user's privacy in a way that the user could not have anticipated. Trust is sticky—that is, if a user trusts a service, they're likely to keep trusting it, and if they don't trust it, they're likely to continue to mistrust it.

Siri is one of many voice assistants that *listen* for spoken commands. The voice assistant recognizes phrases and processes the information. Consider the example described in the car full of kids with the mom engaging a voice feature. In the early days of voice assistants, the AI system listened for simple grammar that took a verb + subject and turned it into a command, like, "Call [Dad]." As the technology continued to improve, dictation of speech to text became more and more accurate.

Definition A **voice assistant (or virtual assistant)** is an AI-based program that allows users to interact with an application through a natural language interface. Virtual assistants have tens of thousands of applications available to users that can perform all manner of tasks for the user: get weather, make cat sounds, tell jokes, sing songs, etc.

When Siri was first released in 2011 on iPhones (in what Apple termed a "beta" version),[12] early reviews were met with cheer and accolades (finally!) there was a voice-based interface. Unfortunately, the honeymoon did not last long. Users began to express frustration and strong negative associations formed. As examples, Siri had issues with both usability and functionality; Siri could be triggered accidentally and its speech recognition was not nearly as robust as users expected. All too often Siri would apologize and say, "Sorry, I don't understand..." And even if it did properly recognize the speech, it frequently misconstrued the request. Very quickly, users also wanted to do more with it than it was designed for. In short, users (err, customers) originally had in their minds an application with such great promise, but the reality was exceedingly modest. Apple had a failure on their hands.[13]

TRUST IS FORMED BY MORE POSITIVE EXPERIENCES

BOB: For any product, whether it has AI or not, the bare minimum should be that it be usable and useful. It needs to be easy to operate, perform the tasks that users ask of it accurately, and not perform tasks it isn't asked to do. That is setting the bar really low, but there are many products in the marketplace that are so poorly designed where this minimum bar is not met.

GAVIN: Just think about how many TV remotes sit on the coffee table of your living room and the sheer number of buttons! It makes you question how much effort went into the design of the remote. Consider the experience of using the remote. It is often at night.

BOB: The worst is the remote that controls the TV settings. Sometimes we fumble in the dimly lit room and accidentally press the wrong button on the remote. When you fumble in the dark, how often has a MENU or SETUP popup window appeared when you thought you pressed the BACK button?

GAVIN: The design challenge becomes even more difficult when we move to AI. Think about the Siri experience which is entirely independent of a screen interface like one on a TV. Because everything happens by voice, the dialog needs to be well developed. It must work or people will abandon it.

---

[12]Tsukayama, Hayley. "Apple's Siri shows she's only a beta." The Washington Post. Last updated November 4, 2011. Accessed June 2, 2020. www.washingtonpost.com/business/technology/apples-siri-shows-shes-only-a-beta/2011/11/04/gIQA6wd-zlM_story.html?
[13]Kinsella, Brett. "How Siri Got Off Track – The Information." Last updated March 14, 2018. Accessed August 14, 2019. https://voicebot.ai/2018/03/14/siri-got-off-track-information/.

BOB: When things work well or the experience is good one, one forms a feeling of trust. And when the voice dialog on Siri does not work, what happens next? Repeat the phrase? But, how many times would someone try? This delivers a feeling of mistrust.

GAVIN: This is important because it can explain why Siri has fallen into disuse. After multiple failures, trust evaporates. The result is that people just stop using the product.

BOB: And with a voice assistant, the feeling can be persistent. Let's say Apple made Siri better and solved some of the "I am sorry, I don't know that yet" interactions. How would you know?

GAVIN: All that effort put in to make Siri smarter would be for naught. This puts the product on a downward path where it is difficult to recover.

The point Our perception of a product is the sum total of the experiences that we have with that product. Does the product deliver the value we had hoped? Our willingness to "trust" the product hangs in that balance.

The role of trust in UX

The field of behavioral economics, which combines psychology and economics, has provided insights that are critical to the pillar of "trust" in UX. Daniel Kahneman, a Nobel Prize winner and critical figure in behavioral economics, divides the brain into two systems.[14] "System 1" is the system of passions that guides the brain: it offers the intuitive judgments that we make constantly and governs our emotional responses to situations. "System 2" is the system of reason: it comes to considered judgments after long periods of analysis. Kahneman wishes to subvert the conventional wisdom that rational thinking is always superior to emotion and that poor decisions usually come from following instincts rather than reason. Kahneman points out that intuition is often effective. System 1 is what allows us to drive a car, maintain most of our social relationships, and often even to answer intellectual questions.

---

[14]Bhalla, Jag. "Kahneman's Mind-Clarifying Strangers: System 1 & System 2." Big Think. Last updated March 7, 2014. Accessed June 2, 2020. https://bigthink.com/errors-we-live-by/kahnemans-mind-clarifying-biases.

Behavioral economists like Kahneman have proposed three important heuristics, or mental shortcuts, that are used often by System 1: affect, availability, and representativeness. For our purposes, we are going to focus on the affect heuristic. The affect heuristic dictates that our initial emotional judgments of someone or something will dictate whether or not we trust that person or thing.[15] This is what sank Siri. Initially, the virtual assistant was cumbersome to use and that affective association of negative feelings with Siri lingered even after the service itself was improved.

In 2014, Amazon came out with its Echo, featuring the Alexa virtual assistant. The device presented the virtual assistant with a particular use scenario in mind—one that was particularly accommodating to virtual assistant use. In the home, where the Amazon Echo is meant to sit, you're less likely to feel ashamed about talking to your tech. That's not the only difference between the Echo and Siri—the Echo also looks completely different, sitting as it does in a cylindrical device that is dedicated entirely to its use.

Amazon worked hard to make sure that its virtual assistant, Alexa, would provide a good experience from its inception. They may have been inspired by the failure of Amazon's Fire Phone just a couple years earlier. The Fire Phone seemed like a minimum viable product,[16] and it flopped out of the gate. But the Echo was different. While building the Echo, Amazon ran tests including the "Wizard of Oz test." In that test, users asked questions, which were fed to a programmer in the next room, who typed in responses that were then played in Alexa's voice.[17] Amazon used the test to analyze the vocal attributes that got the best response from users. Amazon took the time and effort to build a product that would engender trust, and it showed. We don't exactly know what user research Apple did with Siri, but whatever they did, the outcome was not as successful as it was for Amazon.

The point   Trust is vitally important to user adoption, and it's easily lost. Developers need to be careful to design an experience that engenders trust.

---

[15]"Affect heuristic." Behavioral Economics. Accessed June 2, 2020. www.behavioraleco-nomics.com/resources/mini-encyclopedia-of-be/affect-heuristic/.

[16]A "minimum viable product" (MVP) is one that has only the necessary features to deliver value to users and capture market share; frequently, it is also used to learn from the market how to grow and improve the product.

[17]Kim, Eugene. "The inside story of how Amazon created Echo, the next billion-dollar business no one saw coming." Business Insider Australia. Last updated April 2, 2016. Accessed June 2, 2020. www.businessinsider.com.au/the-inside-story-of-how-amazon-created-echo-2016-4.

The need for UX design

At the core of the UX outlook on design is the concept of "affordances," developed by psychologist James Gibson. **Affordances** are points of interaction between object and perceiver that allow the perceiver to understand the object's features (the things that the object can do for the perceiver and for other agents).[18] Gibson sees these properties as extant in the universe.

Certain affordances are easy for us to discover, perhaps directed by cultural norms for using an object or by the object's design. Doors with flat plates afford pushing, while doors with loop handles afford pulling. Google's homepage is dominated by a single text box with a search button and lots of white space, indicating that it allows you to search for anything you'd like. Recognition of the power affordances have to provide information to the user is woven into the design of the product to improve usability, function, and use.

However, some of an object's features may be less clear—meaning that the corresponding affordances are only present for users who are in the know. For tech products, these are the kind of features that end up being revealed to users by accident or through viral online articles. (They have titles like "10 Things You Didn't Know Your Phone Could Do.") If the user does not know that the object has a certain function made clear by an affordance, that function becomes much less useful. Whenever the number of functions exceeds the number of affordances, there is trouble. Therefore, the designer must be skilled at communicating to the user what the object is capable of— in other words, the designer must create "signifiers" (another term from Norman) that communicate the object's affordances (e.g., Google's search box).[19]

Affordance generation is two-sided. An object must have certain properties, and a user must recognize possible functions for those properties. For this reason, users of products may discover affordances that the designer never intended. For example, Facebook likely intended its groups feature to link real-life groups of friends, coworkers, and classmates, but many users have instead used them to share memes and inside jokes with like-minded strangers.

---

[18]Gibson, James J. "The Theory of Affordances." Semantic Scholar. Accessed June 2, 2020. From *The Ecological Approach to Visual Perception*. Houghton Mifflin (Boston): 1979. https://pdfs.semanticscholar.org/eab2/b1523b942ca7ae44e7495c496b-c87628f9e1.pdf.

[19]Norman, Don. "Signifiers, not affordances." jnd.org. Last updated November 17, 2008. Accessed June 2, 2020. https://jnd.org/signifiers_not_affordances/.

Facebook seems to have welcomed the opportunity to retain younger users, even rolling out a new screening feature that particularly helps meme-based groups.[20] After users discovered a new affordance for Facebook's groups feature, Facebook updated its product to reflect a use case that they were never planning for. This is illustrative of the active role that users can play in further shaping the design. Design needs to recognize the many ways that a user might use a feature.

IT'S NOT ALWAYS ABOUT AESTHETICS

GAVIN: Sometimes the user's needs can come in conflict with the aspirations of the designer. Consider Apple's $5 billion dollar state-of-the-art headquarters built in 2018 by architect Norman Foster. The building used rounded glass that was designed to "achieve an exact level of transparency and whiteness." [21]

BOB: The problem was that people could not tell where the door ended and the wall began. And even the building inspector cautioned on this risk. But to the architect, it was all about the design and not about those affordances.

GAVIN: What happened? Workers walked right into the glass so hard that 911 was called three times in the first month! Employees were so fearful that they placed their own affordances on the walls—sticky notes—to prevent more injuries.

BOB: But the building designers removed the sticky notes because it was said to have detracted from the building's design aesthetics.

GAVIN: Not only is this ironic that it happened at Apple, but talk about architects not living in the places they design. We've heard that Apple-approved stickers were made after that to provide better affordances to distracted walkers in an attempt to reduce 911 calls due to injuries.

The point   Sometimes the design for design sake can get in the way. How users actually engage (or in this case, walk) can often be at odds with the aesthetic. Design needs to work for the whole user, not simply what the eye sees.

---

[20]Sung, Morgan. "The only good thing left on Facebook is private meme groups." Mashable. Last updated August 9, 2018. Accessed June 2, 2020. https://mashable.com/article/weird-facebook-specific-meme-groups/.

[21]Gibbs, Samuel (2018). "Is Alexa always listening? New study examines accidental triggers of digital assistants." The Guardian. Last updated March 5, 2018. Accessed June 2, 2020. www.theguardian.com/technology/2018/mar/05/apple-park-workers-hurt-glass-walls-norman-foster-steve-jobs.

The user-centered design ethos that is at the core of UX differs from the stereotypical association of the term "design" with form and aesthetics. While form and aesthetics are certainly important components of an experience, they need to be combined with functionality in order to deliver the best possible experience to the user. UX design focuses on the ways in which form and function can complement one another, without compromising either one *for* the other.

This vision of design was well articulated in a lengthy opinion piece co-authored by Don Norman and Bruce Tognazzini in 2015, criticizing the design of Apple's operating system for its smartphones and tablets.[22] Norman and Tognazzini, who had both worked for Apple during the pre-iDevice days, felt that Apple had once been a leader in user-centered design, but that it had since lost its compass. They centered their criticisms around iOS's lack of certain useful affordances, such as a universal back button to undo actions, as well as its lack of signifiers for many of the affordances it does have.

Apple's gestural interfaces rely on the concept of second nature. Human beings are built to learn and adopt new systems of interacting with the world and quickly become so familiar with them that they become instinctive.[23] This is what we have all done with swiping around our phones, pinching to zoom in and out, and the rest of the gestures that allow us to use smartphones and tablets. Apple, however, has gone the extra mile with its gestural interfaces, working in all manner of different gestures. To see this in action, all you have to do is go to an Apple store and swipe all of the various devices there with three fingers or your palm in different directions and patterns. Odds are the device will start to do lots of unexpected functions.

The problem that Norman and Tognazzini identify is that most users have no natural way of discovering these gestural features. There are no on-screen indicators that these features are present, and very few users are going to experiment with the OS or read the manual in order to find out about them. So, for all intents and purposes, these features don't exist for most users— except to confuse them when they accidentally trigger them while trying to do something else. That leads to negative interactions.

---

[22]Norman, Don and Tognazzini, Bruce. "How Apple Is Giving Design A Bad Name." Fast Company. Last updated November 10, 2015. Accessed June 2, 2020. www.fastcompany. com/3053406/how-apple-is-giving-design-a-bad-name.

[23]Bhalla, Jag. "Inheriting Second Natures." Scientific American. Last updated April 25, 2013. Accessed June 2, 2020. https://blogs.scientificamerican.com/guest-blog/ inheriting-second-natures/.

Users' perception of their experiences is vital to their continuing to return to those same experiences time and time again. People will generally tend to build patterns of engagement with objects in their world that they consider to be profitable and enjoyable.

UX DESCRIBES THE HOLISTIC EXPERIENCE OF INTERACTING WITH A PRODUCT

BOB: If I need to get something done, I'm not going to use an iPad. I'm not going to use a smartphone. I'm certainly not going to tell Alexa. I'm going to turn on my computer and point and click my way through a complicated task. And that's not just because it has a faster processor; it's because a computer is easier to use with respect to complicated tasks.

GAVIN: Norman and Tognazzini pointed it out. When Xerox and Apple were working on the first point-and-click graphical user interfaces, they had UX principles in mind—even if the term wasn't invented yet.

BOB: Today, with touch interfaces, it seems like that's backward. The technology that enabled multi-touch interactions seemed to be "natural gestures" as Steve Jobs called them. In many ways, the gesture itself took precedence over the function. There was almost a look of disgust at those who did not know how to pinch, swipe, or flick with their fingers to interact with the iPhone. It was as if the gesture was more important than the function itself.

GAVIN: Apple's first commercials on the iPhone—which were incidentally paid by AT&T in the US launch in 2007—were all about how to use the iPhone. It was as if Apple made the purpose of their commercials to show the user manual. This was a phenomenal sleight of hand. Who spends hundreds of millions in marketing to show people how to use a product? And Apple argued that touch was so simple—but what came first, the user-manual commercials or the gesture?

BOB: This is a critical consideration as we move into a new era of AI. It seems like we didn't really master how to maximize functionality in touch interfaces to the extent that we did with point-and-click interfaces. But while we try to fix that, we're going to have to turn some of our attention elsewhere. AI brings up all types of new interface possibilities. Voice interfaces are the obvious one, thanks to the virtual assistant. But there's more. There are gestural interfaces where there is a camera watching for movement, like raising your hand in front of smart TV or like the Microsoft Kinect. Computers are starting to read facial expressions and detect affect—even being present in a room is data to AI. Down the road, there may even be neural interfaces—brain waves going directly from your brain to the computer.

BOB: Neural interfaces are a ways away. But I get what you mean, especially with voice. Voice interfaces are a difficult case for usability, even more so than touch, because there are fewer opportunities to communicate affordances to the user. Without any visual signifiers, UX gets a lot harder. With screen-based interfaces, as a designer you can provide visual affordances and rely on people's ability to recognize information. With voice, the interface is natural language—appearing to be wide open to the infinite number of sentences and questions that I could ask.

GAVIN: These are important questions that we're facing in tech right now. But UX people aren't always in the room where the decision makers are, and, more often than not, they should be.

The point   UX describes the holistic experience of interacting with a product.

Conclusion: Where we're going

In the next chapter, we will look at the simultaneous, but independent development of the fields of AI and UX (then called human-computer interaction or human factors) and consider the relevant historical and intersectional points that help us to glean lessons from the past and move toward better design. We'll also discuss the legacy of a few psychologists and how their work shaped both UX and AI. In Chapter 3, we will examine the state of AI today and where UX is and isn't coming into play. We will also take a look at some of the psychological principles underlying human-computer interaction, a key component of interaction. In the later chapters, we will propose, and justify, our UX framework for AI success and discuss its implications for the future.

AI and UX: Parallel Journeys

In this chapter, we're going to take you through some key milestones of both AI and UX, pointing out lessons we take from the formation of the two fields. While the histories of AI and UX can fill entire volumes of their own, we will focus on specific portions of each.

If we step back and look at how AI and UX started as separate disciplines, following those journeys provides an interesting perspective of lessons learned and insight. We believe that at the confluence of these two disciplines is where AI will have much more success.

UX is a relatively modern discipline with its roots in the field of psychology; as technology emerged, it became known as the field of human-computer interaction (HCI). HCI is about optimizing the experience people have with technology. It is about design and it recognizes that a designer's initial attempt on the design of the product might require modification to make the experience a positive one. (We'll discuss this in more detail later in this chapter.) As such, HCI emphasizes an interactive process. At every step of the interaction, there can and should be opportunities for the computer and the human to step back and provide each other feedback, to make sure that each party contributes positively and works comfortably with the other. AI opens up many possibilities in this type of interaction, as AI-enabled computers are becoming capable of learning about humans as a user in the same way that a real flesh-and-blood personal assistant might. This would make a better-calibrated AI assistant, one far more valuable than simply a tool to be operated: a partner rather than a servant.

© Gavin Lew, Robert M. Schumacher 2020
G. Lew and R. M. Schumacher, *AI and UX*,
https://doi.org/10.1007/978-1-4842-5775-3_2

The Turing Test and its impact on AI

The exact start of AI is a subject of discussion, but for practical purposes, we choose to start with the work of computer scientist, Alan Turing. In 1950, Turing proposed a test to determine whether a computer can be said to be acting intelligently. He felt that an intelligent computer was one that could be mistaken for a human being by another human being. His experiment was manifest in several forms that would test for computer intelligence. The clearest form involved a user sending a question to an unknown respondent, either a human or a computer, which would provide an answer anonymously. The user would then be tasked with determining whether this answer came from a human or from a computer. If the user could not identify the category of the respondent with at least 50% accuracy, the computer would be said to have achieved intelligence, thereby passing the "Turing Test."

■ **Definition** The **Turing Test** is a procedure intended to determine the intelligence of a computer by asking a series of questions to assess whether a human is unable to distinguish if a computer or human is giving responses.[1]

The Turing Test has become a defining measure of AI, particularly for AI opponents, who believe that today's AI cannot be said to truly be "intelligent." However, some AI opponents, such as philosopher John Searle, have proposed that Turing's classification of machines that seem human as intelligent may have even gone too far, since Turing's definition of intelligent computers would be limited to machines that imitate humans.[2] Searle argues that intention is missing from the Turing Test and the definition of AI goes beyond syntax.[3] Elon Musk takes a similar view of intelligence[4] to Searle, proposing that AI simply delegates tasks to constituent algorithms which consider individual factors and does not have the ability to consider complex variables on its own, so this would argue AI is not really intelligent at all.

---

[1]Mifsud, Courtney (2017). "A brief history of artificial intelligence." Artificial Intelligence: The Future of Humankind. Time Inc. Books. Pp 20–21.

[2]Searle, John (1980). "Minds, Brains and Programs," The Behavioral and Brain Sciences. 3, pp. 417–424.

[3]Günther, Marios (2012). "Could a machine think? Alan M. Turing vs. John R. Searle." Universite Paris IV Unite de Formation et de Philosphie et Sociologie. January 2012. Accessed June 16, 2020. https://philarchive.org/archive/MARCAM-4.

[4]Gershghorn, Dave. "Elon Musk and Mark Zuckerberg can't agree on what AI is, because no one knows what the term really means." Quartz. Last updated March 30, 2017. Accessed June, 16, 2020. https://qz.com/945102/elon-musk-and-mark-zuckerberg-cant-agree-on-what-ai-is-because-nobody-knows-what-the-term-really-means/.

As far as we know, no computer has ever passed the Turing Test—though a recent demonstration by Google Duplex (making a haircut appointment) was eerily close.[5] The Google Duplex demonstrations are fascinating as they represent examples of natural language dialog. The recordings start with Google's Voice AI placing a telephone call to a human receptionist to schedule a hair appointment and a call to make a reservation with a human hostess at a restaurant.[6] What is fascinating is the verbal and nonverbal cues that were designed into the computer voice, such as pauses and inflection by Duplex, were interpreted by the human successfully. On face, Duplex engaged into a conversation with a human where the human does not appear to realize that a machine is operating at the other end of the call. It's unclear how many iterations Google actually had to go through to get this example. But, in this demonstration, the machine—through both verbal and nonverbal cues— seemed to successfully navigate a human conversation without the human showing any knowledge of or negative reaction to the fact that it was a machine that was calling.

AI HAS A DISTINCTLY HUMAN ELEMENT

BOB: Regardless of whether or not the Turing Test is a sufficient litmus test for the presence of AI, it's had a profound influence on how we define AI.

GAVIN: The Turing Test has captured the attention of the masses interested in the future of AI. The Turing Test has an inherent simplicity as a game of imitation and asked the question: can a computer fool a human?

BOB: Popular science fiction elevated the Turing Test in movies, such as *Ex Machina*, which was explicit in its use of the Turing Test, and *Blade Runner*, where the protagonist played by Harrison Ford tests whether a woman is a human or "replicant."

GAVIN: Turing's influence on computing and our perception of intelligence is undeniable, that's for sure.

The point   Throughout the short history of AI, there has been a desire to make the AI have human qualities.

---

[5]Oppermann, Artem. "Did Google Duplex beat the Turing Test? Yes and No." TowardsDataScience.com. May 20, 2018. Accessed June 16, 2020. https://towardsda-tascience.com/did-google-duplex-beat-the-turing-test-yes-and-no-a2b87d1c9f58. Accessed June 16, 2020.
[6]Leviathan, Yaniv & Matias, Yossi (2018). "Google Duplex: An AI system for accomplishing real-world tasks over the phone." May 8, 2018. Accessed June 16, 2020. https://ai.googleblog.com/2018/05/duplex-ai-system-for-natural-conversation.html. May 8, 2018. Accessed June 16, 2020.

As AI continues to evolve, one trap that needs to be avoided is to advance a definition of AI that juxtaposes AI *against* a human. This might be too limiting to where AI that complements human decision making would be undervalued. Put simply, there is more to AI than conversation. Moreover, by simply defining intelligence based on Turing's test, the development of AI could lead to a belief that a program seems more intelligent than it actually is.

Attribution of intelligence to AI

One of the most notorious programs in computer science was written in 1966 by Joseph Weizenbaum, a professor at MIT. The program was called ELIZA. The name comes from George Bernard Shaw's play, *Pygmalion*, in which Eliza Doolittle studies elocution and etiquette. In this program, Weizenbaum modeled the interaction similar to how a Rogerian psychotherapist might respond to a patient.

The person simply types responses to the questions Eliza asked.

EXAMPLE OF ELIZA CONVERSING WITH A PERSON

ELIZA: Tell me your problem.

PERSON: I am depressed.

ELIZA: How long have you been feeling depressed?

PERSON: Since my daughter left home.

ELIZA: Please go on.

PERSON: I don't want her to get into trouble.

ELIZA: What would it mean to you if she got into trouble?

And so on…

The point   This conversation between the user and the program left an impression that Eliza embodied human-like intelligence. Because the dialog felt natural, the computer must be intelligent. But, did this constitute intelligence? Is all that is necessary for intelligence is to fool a human?

ELIZA was successful and people may have poured their lives into it. But, in all of these conversations, there was no learning algorithm to carefully analyze the data. In fact, not much of what was typed was saved as this was 1966. Some proclaimed that Weizenbaum had solved natural language through his program.

Weizenbaum ended up on a crusade against his own program.[7] ELIZA was more of a Carl Rogers parody. The program did not know psychology, just semantic logic to reflect questions back. But because the program *felt* human, intelligence was bestowed to it. This is an example of AI that can capture the imagination of the masses. And this makes AI susceptible to being overhyped.

The influence of hype

With ELIZA, the hype was brought about from the users. In other examples, hype can come from its creators, investors, government, media, or market forces.

Historically, AI overpromised and under-delivered

In the past, artificial intelligence has had something of an image problem. In 2006, *The New York Times*' John Markoff called AI "a technology field that for decades has overpromised and under-delivered"[8] in the lead paragraph of a story about an AI success.

One of the earliest attempts to develop AI was machine translation, which has its genesis in the post-World War II information theories of Claude Shannon and Norbert Weaver; there was substantial progress in code breaking as well as theories about universal principles underlying language.[9]

Definition **Machine translation** is the translation of one language into another via a computer program.

Machine translation's paramount example is the now (in)famous Georgetown-IBM experiment[10] where in a public demonstration in 1954, a program developed by Georgetown University and IBM researchers successfully

---

[7]Campbell-Kelly, Martin (2008). "Professor Joseph Weizenbaum: Creator of the 'Eliza' program." The Independent. Independent News and Media. March 18, 2008. Accessed June 16, 2020. www.independent.co.uk/news/obituaries/professor-joseph-weizenbaum-creator-of-the-eliza-program-797162.html.

[8]Markoff, John. "Behind Artificial Intelligence, a Squadron of Bright Real People." New York Times. October 14, 2005. Accessed June 16, 2020. www.nytimes.com/2005/10/14/technology/behind-artificial-intelligence-a-squadron-of-bright-real-people.html.

[9]Hutchins, W. John. "The history of machine translation in a nutshell." Last updated November 2005. Accessed June 16, 2020. www.hutchinsweb.me.uk/Nutshell-2005.pdf.

[10]Hutchins, W. John. "The Georgetown-IBM Experiment Demonstrated in January 1954." In *Conference of the Association for Machine Translation in the Americas*, pp. 102–114. Springer, Berlin, Heidelberg, 2004.

translated many Russian-language sentences to English. This demonstration earned the experiment major media coverage. In the heat of the Cold War, a machine that could translate Russian documents into English would have been very compelling to American national defense interests. A number of headlines—"The bilingual machine," for example—greatly exaggerated the machine's capabilities.[11] This coverage was accompanied by massive investment leading to wild predictions about the future capabilities of machine translation. One professor who worked on the experiment was quoted in the *Christian Science Monitor* saying that machine translation in "important functional areas of several areas" might be ready in 3–5 years.[12] Hype was running extremely high. The reality was far different: the machine could only translate 250 words and 49 sentences.

Indeed, the program's focus was on translating a set of narrow scientific sentences in the domain of chemistry, but the press coverage focused more on a select group of "less specific" examples which were included with the experiment. According to linguist W. John Hutchins,[13] even these few less specific examples shared features in common with the scientific sentences that made them easier for the system to analyze. Perhaps because of these few contrary examples, the people covering the Georgetown-IBM experiment did not grasp the leap in difficulty between translating a defined set of static sentences to translating something as complex and dynamic as policy documents or newspapers.

The Georgetown-IBM translator may have seemed intelligent in its initial testing, but further analysis proved its limitations. For one thing, it was based on a rigid rules-based system. Just six rules were used to encode the entire conversion from English to Russian.[14] This, obviously, inadequately captures the complexity of the task of translation. Plus, language only loosely follows rules—for proof, look no further than the plethora of irregular verbs in any language.[15] Not to belabor the point but the program was trained on a narrow corpus and its main function was to translate scientific sentences, which is only an initial step towards translating Russian documents and communications.

This very early public test of **machine translation** featured AI that seemed to pass the Turing Test—but that accomplishment was deceptive.

---

[11]Hutchins, "Georgetown," 103.

[12]Hutchins, "Georgetown," 104.

[13]Hutchins, "Georgetown," 112.

[14]Hutchins, "Georgetown," 106–107.

[15]For instance, the verb "to walk" follows a regular structure: I walk, you walk, she walks, we walk, and they walk. The verb "to be" however is highly irregular: I am, you are, he is, we are, and they are. In many languages, high-frequency verbs are irregular in their construction and usage.

DON'T BELIEVE THAT THE POWER OF
COMPUTING CAN OVERCOME ALL

GAVIN: The Georgetown-IBM experiment was a machine language initiative that started as a demonstration to translate certain chemistry documents. And this resulted in investment that spurred a decade of research in machine language.

BOB: Looking back now, you could argue the logic of applying something that marginally worked in the domain chemistry to be generalized to the entire Russian language. It seems overly simplistic, but, at the time, this chemistry corpus of terms might have been the best available dataset. Over the past 70 years, the field of linguistics has evolved significantly, and the nuisances of language are now recognized to be far more complex.

GAVIN: Nevertheless, the fascination that the power of computing would find patterns and solve mutual translation between English and Russian is an all too common theme. I suspect that the researchers were clear in the limitations of the work, but as with Greenspan's now famous "irrational exuberance" quote that described the hype associated with the stock market, expectations can often take on a life of their own.

The point   We must not believe that the power of computing can overcome all. What shows promise in one domain (chemistry) might not be widely generalizable to others.

The Georgetown and IBM researchers who presented the program in public may have chosen to hide the flaws of their machine translator. They did so by limiting the translation to the scientific sentences that the machine could handle. The few selected sentences that the machine translated during the demonstration were likely chosen to fit into the tightly constrained rules and vocabulary of the system.[16]

The weakness of the Turing Test as a measure of intelligence can be seen in journalists' and funders' initial, hype-building reactions to the Georgetown-IBM experiment's deceptively human-like results.[17] Upon witnessing a machine that could seemingly translate Russian sentences with the near accuracy of a human translator, journalists[18] must have thought they had seen something whose capabilities significantly outstripped the reality of the program.

Yet these journalists were ignorant or unaware of the limited nature of the Georgetown-IBM technology (and the organizers of the experiment may have nudged them in that direction with their choices for public display). If the

---

[16]Hutchins, "Georgetown," 110.
[17]Some tens of millions of dollars were provided by funding agencies.
[18]Hutchins, "Georgetown," 103.

machine had been tested on sentences outside the few that were preselected by the researchers, it wouldn't have appeared to be so impressive. Journalists wrote articles that hyped the technology's capability. But the technology wasn't ready to match the hype. Nearly 60 years later, machine translation is still considered imperfect at best.[19]

The point   Hype can play a large influence on whether the product is judged a success or failure.

AI failures resulted in AI winters

A most devastating consequence of this irrational hype was the suspension of funding for AI research. As described, the hype and perceived success from the Georgetown-IBM experiment resulted in massive interest and substantially increased investment in machine translation research; however, that research soon stagnated as the difficulty of the real challenge associated with **machine translation** began to sink in.[20] By the late 1960s, the bloom had come off the rose. Hutchins specifically tied funding cuts to the Automatic Language Processing Advisory Committee (ALPAC) report, released in 1966.[21]

The ALPAC report, sponsored by several US government agencies in science and national security, was highly critical of machine translation, implying that it was less efficient and more costly than human-based translation for the task of translating Russian documents.[22] At best, the report said that computing could be a tool for use in human translation and linguistics studies, but not as a translator itself.[23] The report went on to say that machine-translated text needed further editing from human translators, which seemed to defeat the purpose of using it in place of human translators.[24] The conclusions of the report led to a drastic reduction in machine translation funding for many years afterward.

In a key portion, the report used the Georgetown-IBM experiment as evidence that **machine translation** had not improved in a decade's worth of effort. The report compared the Georgetown-IBM results directly with results from

---

[19]"Will Machines Ever Master Translation?" IEEE Spectrum. Last updated January 15, 2013. Accessed June 16, 2020. https://spectrum.ieee.org/podcast/robotics/artifi-cial-intelligence/will-machines-ever-master-translation.

[20]Hutchins, "Georgetown," 113.

[21]Hutchins, John. "ALPAC: the (in)famous report." Originally published in *MT News International* 14 (1996). Accessed June 16, 2020. www.hutchinsweb.me.uk/ALPAC-1996.pdf.

[22]Hutchins, "ALPAC," 2, 6.

[23]Hutchins, "ALPAC," 6.

[24]Hutchins, "ALPAC," 3.

subsequent Georgetown machine translators, finding that original Georgetown-IBM's results had been more accurate than advanced versions. That said, Hutchins defined the original Georgetown-IBM experiment as not an authentic test of the latest machine translation technology but as a spectacle "intended to generate attention and funds."[25] Despite this, ALPAC judged later results against Georgetown-IBM as if it had been a true showing of AI's capabilities. Even though **machine translation** may have actually improved in the late 1950s and early 1960s, it was judged against its hype, not against its capabilities.

As machine translation was one of the most important early manifestations of AI, this report had an impact on the field of AI in general. The ALPAC report and the corresponding domain-specific machine translation winter were part of a chain reaction that eventually led to what is considered the first **AI winter**.[26]

Definition   An **AI winter** is a period when research and investment into AI stagnates significantly. During these periods, AI development gains a negative reputation as an intractable problem. This leads to decreased investment in AI research, which further exacerbates the problem. We identify two types of AI winters: some are domain specific, where only a certain subfield of AI is affected, and some are general, in which the entire field of AI research is affected.

Today, there are lots of different terms for technology that encapsulate AI—expert systems, machine learning, neural networks, deep learning, chatbots, and many more. Much of that renaming started in the 1970s, when AI became a bad word. After early developments in AI in the 1950s, the field was hot—not too different from right now, though on a smaller scale. But in the decade or two that followed, funding agencies (specifically US and UK governments) labeled the work a failure and halted funding—the first-ever general AI winter.[27]

AI suffered greatly from this long-term lapse in funding. In order to get around AI's newfound negative reputation, AI researchers had to come up with new terms that specifically did not mention AI to get funding. So, following the AI winter, new labels like *expert systems* emerged.

---

[25]Hutchins, 113.

[26]Bostrom, Nick. *Superintelligence: Paths, Dangers, Strategies*. New York: Oxford University Press, 2014. 8.

[27]Schmelzer, Ron (2018). "Are we heading into another AI winter?" Medium. Posted June 26, 2018. Accessed September 4, 2019. https://medium.com/cognilytica/are-we-heading-to-another-ai-winter-e4e30acb60b2.

Given the seeming promise of AI today, it may be difficult to contemplate that another AI winter may be just over the horizon. While there is much effort and investment directed toward AI, progress in AI has been prone to stagnation and pessimism in the past.

If AI does enter another winter, we believe a significant contributing factor will be that AI designers and developers neglected the role UX plays in successful design. There is another contributing factor and that is the velocity of technology infusion into everyday life. In the 1950s, many homes had no television or telephone. Now the demands are higher for applications; users will reject applications with poor UX. As AI gets embedded into more consumer applications where user expectations are higher, it is only inevitable that AI will need better UX.

The first AI winter followed the ALPAC report and was associated with a governmental stop to funding related to machine language efforts. This investment freeze lasted into the 1970s in the United States. Negative funding attention and news continued with the 1973 Lighthill Report, where Sir James Lighthill reported to English Parliament results similar to those of ALPAC. AI was directly criticized as being overhyped and not delivering on its promise.[28]

AI BY ANY OTHER NAME

BOB: So, was it that the underlying theory and technology in the Georgetown-IBM experiment were flawed or was it just hype that created the failure?

GAVIN: I think it was both. The ALPAC report pulled no punches and led to a collapse in any research in machine translation—a domain-specific AI winter. Huge hype for machine translation turned out to be misplaced and the result was a significant cut in funding.

BOB: Yes, funding requests with the terms "machine translation" or "artificial intelligence" disappeared. Not unlike the old adage of "throwing the baby out with the bath water," a major failure in one domain makes the whole field look suspect. That's the danger of hype. If it doesn't match the actual capabilities of the product, it can be hard to regain the trust.

GAVIN: The first general AI winter formed a pattern of initial signs of promise, to hype, to failure, and subsequently a freeze in future funding. This cycle led to significant consequences for the field. But scientists are smart; out from the ashes, AI bloomed again, but this time using new terminology such as *expert systems,* which led to advancements in robotics.

---

[28]Hendler, James. "Avoiding Another AI Winter." IEEE Intelligent Systems, 2008. Accessed May 15, 2019. www.researchgate.net/publication/3454567_Avoiding_Another_AI_Winter.

BOB: So, under its new names, AI garnered over $1 billion in new investment in the 1980s, ushered in by private sector companies in the United States, Britain, and Japan.

GAVIN: Actually, Japan's advances in AI spawned US and British international competition to keep up with the Japanese. Notable examples are the European Strategic Program on Research and Information Technology, the Strategic Computing Initiative, and Microelectronics and Computer Technology Corporation in the United States. Unfortunately, hype emerged again, and when these companies failed to deliver on the lofty promises, the second AI winter[29] was said to occur in 1993.

The point   AI has seen boom and bust cycles multiple times in its history.

Can another AI winter happen? It already did

Often lessons from the past are ignored with the *hope that this time will be different*. Will another **AI winter** happen in our lifetime is not the question because one happened before our very eyes.

Consider Apple's **voice assistant,** Siri. Siri was not launched fully functional right out of the gates. The "beta" version was introduced with a lot of fanfare. Soon Apple pulled it out of "beta" and released more fully fledged versions in subsequent updates to iOS—versions that were far more functional and usable than the original—the potential for many users to adapt to it was greatly reduced. However, Siri users had already formed their impressions, and considering the **AI-UX principle** of **trust**, those impressions were long-lasting. Not to be too cheeky, but one bad Apple (released too early) spoiled the barrel.

HOW SIRI IMPACTED CORTANA

BOB: Look, to the Siri fans out there, Apple did an amazing job relative to previous **voice assistants**. When working for Baby Bell companies many years back, we often tested **voice assistants**. Siri was a generation ahead of anything we had in our labs.

GAVIN: And Siri was the first-ever **virtual assistant** to achieve major market penetration. In 2016, industry researcher Carolina Milanesi found that 98% of iPhone users had given Siri at least one chance.[30] This is a phenomenal achievement in mass use of a product.

---

[29]Hubbs, Christian (2019). "The dangers of government funded artificial intelligence." Mises Institute. https://mises.org/wire/dangers-government-funded-artificial-intelligence. Posted March 30, 2019. Accessed August 26, 2019.

[30]Milanesi, Carolina. "Voice Assistant Anyone? Yes please, but not in public!" Creative Strategies. Last updated June 16, 2020. Accessed June 26, 2019. https://creativestrategies.com/voice-assistant-anyone-yes-please-but-not-in-public/.

BOB: The problem though was *continued use.* When 98% were asked how much they used it, most replied "rarely" or "sometimes" (70%). In short, almost all *tried it,* but most *stopped using it.*

GAVIN: Apple hyped Siri for its ability to understand the spoken word, and Siri captured the attention of the masses. But after time, most users were sorely disappointed with hearing the response, "I'm sorry. I don't understand that," and abandoned it after a few initial failures.

BOB: To have so many try a product that is designed to be used daily (i.e., "Siri, what is the weather like today?") and practically abandon its use is not simply a shame; it is a commercial loss. The effort to get a customer to try something and lose them, well, you poison the well.

GAVIN: Even now, if you were to play the Siri prompt ("bee boom"), a chill goes up my spine because I must have accidentally pressed it. But this feeling of a chill negatively impacted other **voice assistants**. Ask yourself: Have you ever tried Cortana (Microsoft's voice feature on Windows OS)? Did you try it? Even once? And why did you not try it?

BOB: No. Never gave it a try. Because to me, Cortana was just another Siri. In fact, I moved to Android partly because Siri was so lame.

GAVIN: In speaking to Microsoft Cortana design and development teams, they would vociferously argue how much different (or better) their **voice assistant** Cortana was from Siri. But because of the failure of **trust**, people who used Siri tended to associate the technology with Cortana.

BOB: Ask if anyone has tried Bixby, Samsung's mobile phone **voice assistant**, and you get blank stares.

The point   Violating an **AI-UX principle** like **trust** can be powerful enough to prevent users from trying similar, but competitive products. This is arguably a domain-specific AI winter.

These negative feelings toward Siri extended to other virtual assistants that were perceived to be similar to Siri. As other virtual assistants came along, some users had already generalized their experiences with virtual assistants as a category and reached their own conclusions. The immediate impact of this was to reduce the likelihood of adoption. For instance, only 22% of Windows PC users ended up using Cortana.[31]

---

[31]Bacchus, Arif. "In the age of Alexa and Siri, Cortana's halo has gone dim." Digital Trends. Last updated February 16, 2019. Accessed June 16, 2020. www.digitaltrends.com/computing/cortana-is-dead/.

Ultimately, Cortana was likely hit even harder than Siri itself by this **AI winter** because Siri was able to overcome and still exists. Cortana was eventually repurposed as a lesser service. In 2019, Microsoft announced that, going forward, they intended to make Cortana a "skill" or "app" for users of various virtual assistants and operating systems that would allow them to access information for subscribers to the Microsoft 365 productivity suite.[32] This meant that Cortana would no longer be equivalent to Siri.

Unlike Siri, Cortana was a vastly capable **virtual assistant** at its launch, especially for productivity functions. Its "notebook" feature, modeled after the notebooks that human personal assistants keep on their clients' idiosyncrasies, offered an unmatched level of personalization.[33] Cortana's notebook also offered users the ability to delete some of the data that it had collected on them. This privacy feature exceeded any offered by other assistants.[34]

Despite these very different capabilities, users simply did not engage. Many could not get past what they thought Siri represented.

Moreover, **interaction** also became a problem for Siri. Speaking to your phone was accompanied by social stigma. In 2016, industry research by Creative Strategies indicated that "shame" about talking to a smartphone in public was a prominent reason why many users did not use Siri regularly.[35] The most stigmatized places for **voice assistant** use—public spaces—also happen to be common use cases for smartphones. Ditto for many of the common use cases for laptop computers: workplace, library, and classroom. Though in our very unscientific observations, an increasing number of people are using the voice recognition services on their phones these days.

---

[32]Warren, Tom. "Microsoft No Longer Sees Cortana as an Alexa or Google Assistant Competitor" The Verge. January 18, 2019. Accessed June 16, 2020. www.theverge.com/2019/1/18/18187992/microsoft-cortana-satya-nadella-alexa-google-assistant-competitor.

[33]Beres, Damon. "Microsoft's Cortana Is Like Siri With A Human Personality." HuffPost. June 29, 2015. Accessed June 16, 2020. www.huffpost.com/entry/microsofts-cortana-is-like-siri-with-a-human-personality_n_55b7be94e4b0a13f9d1a685a.

[34]Hachman, Mark. "Microsoft's Cortana guards user privacy with 'Notebook.'" PC World. Last updated February 21, 2014. Accessed June 16, 2020. www.pcworld.com/article/2099943/microsofts-cortana-digital-assistant-guards-user-privacy-with-notebook.html.

[35]Reisinger, Don. "You're embarrassed to use Siri in public, aren't you?" Fortune. Last updated June 6, 2016. Accessed June 16, 2020. http://fortune.com/2016/06/06/siri-use-public-apple/.

THE EMERGENCE OF ALEXA

BOB: Perhaps the reason we do not readily think of the impact that stemmed from the poor initial **MVP** experience with Siri is because this **AI winter** lasted only a couple years not decades. This rebirth of the **virtual assistant** emerged as Amazon's Alexa.

GAVIN: But look what it took for the masses to try another **voice assistant**. Alexa embodied an entirely new form factor, something that sat like a black obelisk on the kitchen counter. This changed the environment of use. Where the device was placed afforded a visual cue to engage Alexa.

BOB: It also allowed Amazon to bring forth Alexa with more features than Siri. Amazon was determined to learn from the failed experience of its Amazon Fire Phone. The Fire had a **voice assistant** feature, and Amazon's Jeff Bezos did not want to make Alexa's **voice assistant** an **MVP** version. He wanted to think big.

GAVIN: Almost overnight, Jeff Bezos dropped $50 million and authorized headcount of 200 to "build a cloud-based computer that would respond to voice commands, 'like the one in Star Trek'."[36]

The point   Alexa emerged as a **voice assistant** and broke out of the **AI winter** that similar products could not, but it needed an entirely different form factor to get users to try it. And when users tried, Jeff Bezos was determined to not have users experience an MVP version, but much bigger.

"Lick" and the origins of UX

In the early days of computing, computers were seen as a means of extending human capability by making computation faster. In fact, through the 1930s, "computer" was a name used for humans whose job it was to make calculations.[37] But there were a few who saw computers and computing quite differently. The one person who foresaw what computing was to become was J. C. R. Licklider, also known as "Lick." Lick did not start out as a computer scientist; he was an experimental psychologist, to be more precise, a highly regarded psychoacoustician, a psychologist who studies the perception of

---

[36]Bariso, Justin (2019). "Jeff Bezos Gave an Amazon Employee Extraordinary Advice After His Epic Fail. It's a Lesson in Emotional Intelligence. The story of how Amazon turned a spectacular failure into something brilliant." Inc. December 9, 2019. Accessed June 16, 2020. www. inc.com/justin-bariso/jeff-bezos-gave-an-amazon-employee-extraordinary-advice-after-his-epic-fail-its-a-lesson-in-emotional-intelligence.html.

[37]Montecino, Virginia. "History of Computing." George Mason University. Last updated November 2010. Accessed June 16, 2020. https://mason.gmu.edu/~montecin/computer-hist-web.htm.

sound. Lick worked at MIT's Lincoln Labs and started a program in the 1950s to introduce engineering students to psychology—a precursor to future **human-computer interaction** (HCI) university programs.

Definition **Human-computer interaction** is an area of research dedicated to understanding of how people interact with computers and the application of certain psychological principles to the design of computer systems. [38]

Lick became head of MIT's human factors group where he transitioned from work in psychoacoustics to computer science because of his strong belief that digital computers would be best used in tandem with human beings to augment and extend each other's capabilities.[39] In his most well-known paper, Man-Computer Symbiosis[40], Lick described a computer assistant that would answer questions when asked, do simulations, display results in graphical form, and extrapolate solutions for new situations from past experience.[41] (Sounds a little like AI, doesn't it?) He also conceived the "Intergalactic Computer Network" in 1963—an idea that heralded the modern-day Internet.[42]

Eventually, Lick was recognized for his expertise and became the head of the Information Processing Techniques Office (IPTO) of the US Department of Defense Advanced Research Projects Agency (ARPA). Once there, Lick fully embraced his new career in computer engineering. He was given a budget of over $10 million dollars to launch the vision he cited in *Man-Computer Symbiosis*. In the intertwining of HCI and AI, Lick was the one who initially funded the work of the AI and Internet pioneers Marvin Minsky, Douglas Engelbart, Allen Newell, Herb Simon, and John McCarthy.[43] Through this funding, he spawned many of the computing "things" we know today (e.g., the mouse, hypertext, time-shared computing, windows, tablet, etc.). Who could

---

[38]Carroll, John M. & Kjeldskov, J. (2013). "The encyclopedia of human-computer interaction. 2nd Edition." Interaction Design Foundation. www.interaction-design.org/literature/book/the-encyclopedia-of-human-computer-interaction-2nd-ed/human-computer-interaction-brief-intro. Accessed August 26, 2019.

[39]Hafner, Katie and Lyon, Matthew. Where Wizards Stay Up Late: The Origins of the Internet, 10–13, 28–47. New York: Simon & Schuster (1996).

[40]Licklider, J. C. R., "Man-Computer Symbiosis," *IRE Transactions on Human Factors in Electronics*, vol. HFE-1, 4–11, March 1960.

[41]"Joseph Licklider." https://history-computer.com/Internet/Birth/Licklider.html. Retrieved July 30, 2019.

[42]Licklider, J. C. R. (23 April 1963). "Topics for Discussion at the Forthcoming Meeting, Memorandum For: Members and Affiliates of the Intergalactic Computer Network." Washington, D.C.: Advanced Research Projects Agency, via KurzweilAI.net. Retrieved August 18, 2019.

[43]"Joseph Licklider," History-Computer, https://history-computer.com/Internet/Birth/Licklider.html. Accessed July 30, 2019.

have predicted that a humble experimental psychologist, turned computer scientist, would be known as the "Johnny Appleseed of the Internet"?[44]

AI AND HUMANS ARE COMPLEMENTARY

GAVIN: Bob, you're a huge fan of Lick.

BOB: With good reason. Lick was the first person to merge principles of psychology into computer science. His work was foundational for computer science, AI and UX. Lick pioneered an idea essential to UX that computers can and should be leveraged for efficient collaboration among people.

GAVIN: You can certainly see that in technology. Computers have become the primary place where communication and collaboration happen. I can have a digital meeting with someone who's halfway across the world and collaborate with them on a project. It seems obvious to us, but it's really a monumental difference from the way the world was even just 20 years ago, let alone in Lick's day.

BOB: We now exist in a world where computers are not just calculators but the primary medium of communication among humans. That vision came from Lick and others like him who saw the potential of digital technology to facilitate communication.

The point   Lick formed the basis of where AI is headed today—where AI and humans are complementary.

Licklider's legacy lived on in others, particularly of note is Robert (Bob) Taylor. Taylor had been greatly influenced by Lick's ideas in *Man-Computer Symbiosis* and had a similar DNA to Lick's from psychologist to psychoacoustician to computer scientist. Lick and Taylor met in 1962 when Lick was running the IPTO at ARPA. They co-authored a paper in 1968 called "The Computer as a Communication Device,"[45] illustrating their shared vision of using computers to enhance human communication.[46] They begin the paper with the following two sentences:

> *In a few years, men will be able to communicate more effectively through a machine than face to face. That is a rather startling thing to say, but it is our conclusion.*

---

[44]Waldrop, M. Mitchell (2001). *The Dream Machine: J. C. R. Licklider and the Revolution That Made Computing Personal.* New York: Viking Penguin. p. 470.

[45]J. C. R. Licklider; Robert Taylor (April 1968). "The Computer as a Communication Device." *Science and Technology.*

[46]"Robert Taylor." Internet Hall of Fame. Accessed July 9, 2019. www.internethalloff-ame.org/inductees/robert-taylor.

Lick and Taylor describe, in 1968, a future world that must have seemed very odd at the time. Fast forward to today where our lives are filled with video calls, email, text messaging, and social media. This goes to show how differently people thought of computing back in the 1960s and how forward-looking Lick and Taylor were at the time. This paper was a clear-eyed vision of the Internet and how we communicate today.

Taylor eventually succeeded Lick as the director of IPTO. While there, he started development on a networking service that allowed users to access the information stored on remote computers.[47] One of the problems he saw though was that each of the groups he funded were isolated communities and were unable to communicate with one another. His vision to interconnect these communities gave rise to the ARPANET and eventually the Internet.

After finishing his time at IPTO, Taylor eventually found his way to Xerox PARC (Palo Alto Research Center) and managed its Computer Science Lab, a pioneering laboratory for new and developing computing technologies that would go on to change the world as we know it. We'll discuss Xerox PARC later in this chapter. But, first, let's return to the world of AI and see what was going on during this time period.

Expert systems and the second AI winter

Following the first AI winter that was initiated by the ALPAC findings that concluded unfavorable progress in machine translation, scientists eventually adapted and proposed research into new AI concepts. This was the rise of **expert systems** in the late 1970s and into the 1980s. Instead of focusing on translation, an expert system was a type of AI that used rule-based systems to systematically solve problems.[48]

Definition   Expert systems operate based on a set of if-then rules and draw upon a "knowledge base" that mimics, in some way, how experts might perform a task.

According to Edward Feigenbaum, one of the early AI pioneers following the first AI winter, expert systems brought positive impacts of computer science in mathematics and statistics to other, more qualitative fields.[49, 50] In the

---

[47]Hafner, Lyon, Where Wizards Stay Up Late.

[48]Bostrom, *Superintelligence*, 9.

[49]Feigenbaum, Edward A. "Knowledge Engineering: The Applied Side of Artificial Intelligence." No. STAN-CS-80-812. Stanford Heuristics Programming Project, 1980. Accessed May 20, 2019.

[50]Feigenbaum. "Knowledge Engineering." 9.0.

1980s, expert systems had a massive spike in popularity, as they entered popular usage in corporate settings. Though expert systems are still used for business applications and emerge as concepts like *clinical decision making* for electronic health record systems (EHR),[51] their popularity fell dramatically in the late 1980s and early 1990s, as an AI winter hit.[52]

Feigenbaum outlined two components of an expert system: the "knowledge base," a set of if-then rules which includes expert-level formal and informal knowledge in a particular field, and the "inference engine," a system for weighting the information from the knowledge base in order to apply it to particular situations.[53] While many expert systems benefit from machine learning, meaning they can adjust their rules without programmer input, even these adaptable expert systems are generally reliant on the knowledge entered into them, at least as a starting point.

This dependence on programmed rules poses problems when expert systems are applied to highly specific fields of inquiry. Feigenbaum identified such a problem[54] in 1980, citing a "bottleneck" in "knowledge acquisition" that resulted from the difficulty in programming expert knowledge into a computer. Since machine learning was unable to directly translate expert knowledge texts into its knowledge base and since experts in many fields did not have the computer science knowledge necessary to program the expert system themselves, programmers acted as an intermediary between experts and AI. If programmers misinterpreted or misrepresented expert knowledge, the resulting misinformation would become part of the expert system. This was particularly problematic in cases where the experts' knowledge was of the unstated sort that comes with extensive experience within a field. If the expert could not properly express this unstated knowledge, it would be difficult to program it into the expert system. In fact, psychologists tried to get at this problem of "knowledge elicitation" from experts in order to support the development of expert systems.[55] Getting people (particularly experts) to talk about what they know and express that knowledge in rules-based format suitable for machines turns out to be a gnarly problem.

These limitations of the expert system architecture were part of the problem that eventually put them into decline. The failure of expert systems led to a years-long period in which the development of AI in general was stagnant.

---

[51]How Can Artificial Intelligence (AI) Improve Clinician EHR Use? Jason, Christopher. December 2, 2019. https://ehrintelligence.com/news/how-can-artificial-intelligence-ai-improve-clinician-ehr-use. Accessed May 22, 2020).
[52]Bostrom, *Superintelligence*, 9.
[53]Feigenbaum. "Knowledge Engineering," 1.2.
[54]Feigenbaum. "Knowledge Engineering," 10.4.
[55]Hoffman, RR (Ed.). (1992). The psychology of expertise: Cognitive research and empirical AI. New York, NY, US: Springer-Verlag Publishing.

We cannot say exactly why expert systems stalled in the 1980s, although irrationally high expectations for a limited form of AI certainly played a role. But it is likely that the perceived failures of expert systems negatively impacted other areas of AI.

AI BEGAN TO EMBRACE COMPLEXITY

GAVIN: Just think about what it took to build "rule-based" systems. You needed computer scientists who programmed the "brain," but you also needed to enter "information" to essentially embed domain knowledge into the system as data.

BOB: When the objective was **machine translation**, the elements were words and sentences. But when you are building **expert systems** like autonomous robotics, this effort adds a physical dimension, like one that would perform on an automated assembly line.

GAVIN: The sheer amount of knowledge at play makes for a complicated world, one that had some programmers coding. Others worked to take knowledge and create training datasets. Others worked on computer vision. Still others worked on robotic functions to enable the mechanical degrees of freedom to complete physical actions. The need to have machines learn on their own has necessitated our current definitions of artificial intelligence. There was simply too much work to be done.

BOB: AI winters have come and gone—but they were hardly the "Dark Ages." The science advanced. As technology advanced, challenges only became greater. Whether changing its name or its focus, many pushed through AI failures to get us to where we are today.

The point   AI winters stifled funding, but the challenge of AI captured the attention of great minds who wanted to advance technology and science.

Of course, failure provides the lessons that we carry forward when we pick ourselves up, dust ourselves off, and move on. Failure helps us be better prepared for the next time we are at a crossroads. We think that AI is at such a crossroads now and that lessons learned from the failure that led to the expert system AI winter can help us get through it. AI scholar Roger Schank,[56] a contemporary of Feigenbaum and other expert systems proponents, outlined his opinion on the shortcomings of expert systems in 1991. Schank believes that expert systems, especially after encouragement from venture capitalists, were done in by an overemphasized focus on their inference engines.[57]

---

[56]Schank, Roger C. "Where's the AI?" AI Magazine 12/4 (1991): 38–49. Accessed May 21, 2019. www.aaai.org/ojs/index.php/aimagazine/article/view/917/835.
[57]Schank, 40.

Schank describes venture capitalists seeing dollar signs and encouraging the development of an inference machine "shell"—a sort of build-your-own-expert-system machine. They could sell this general engine to various types of companies who would program it with their specific expertise. The problem with this approach, for Schank, is that the inference engine is not really doing much of the work in an expert system.[58] All it does, he says, is choose an output based on values already represented in the knowledge base. Just like machine translation, the inference engine developed hype that was incommensurate with its actual capabilities.

These inference machine "shells" lost the intelligence found in the programmers' learning process. Programmers were constantly learning about the expert knowledge in a particular field and then adding that knowledge into the knowledge base.[59] Since there is no such thing as expertise without a specific domain on which to work, Schank argues that the shells that venture capitalists attempted to create were not AI at all—that is, the AI is in the knowledge base, not the rules engine.

The point   Failure can be devastating, but can teach us valuable lessons.

Xerox PARC and trusting human-centered insights

The history of Xerox's Palo Alto Research Center (PARC) is remarkable that a company known for its copiers gave us some of the greatest innovations of all time. In the last decades of the 20th century, Xerox PARC was the premier tech research facility in the world. Here are just some of the important innovations in which Xerox PARC played a major role: personal computer, graphical user interface (GUI), laser printer, computer mouse, object-oriented programming (Smalltalk), and Ethernet.[60] The GUI and the mouse made computing much easier for most people to understand, allowing them to use the computer's capabilities without having to learn complex commands. The design of the early systems was made easier by applying psychological principles to how computers—both software and hardware—were designed.

---

[58]Schank, 45.
[59]Schank, 45.
[60]Dennis, "Xerox PARC."

TECH'S GREATEST ACCOMPLISHMENTS HAD PSYCHOLOGICAL ROOTS

BOB: Bob Taylor, who was head of the Computer Sciences Division at Xerox PARC, recruited the brightest minds from his ARPA network and other Bay Area institutions, such as Douglas Engelbart's Augmentation Research Center. These scientists introduced the concepts of the computer mouse, windows-based interfaces, and networking.

GAVIN: Xerox PARC was one of those places that had a center of excellence (*COE*) that attracted the world's brightest. This wasn't like the COEs we see today that are used as a business strategy. Instead, PARC was recognized like Niels Bohr's institute at Copenhagen when it was the world center for quantum physics in the 1920s or the way postwar Greenwich Village drew artists inspired by Abstract Expressionism or how Motown Records attracted the most creative writers and musicians in soul music.[61] It was vibrant!

BOB: PARC was indeed an institute of knowledge. Despite building such a remarkable combination of talent and innovative new ideas, the sustainability of such an institution can still be transitory. By the 1980s, the diffusion of Xerox PARC scientists began. But much of where technology stands today is because of Xerox PARC's gathering and the eventual dispersion that allowed advancement to move from invention and research to commercialization.

The point   Xerox PARC is where human-computer interaction made progress in technology with roots in psychology.

Eric Schmidt, former chairman of Google and later Alphabet, said—perhaps with a bit of exaggeration—that "Bob Taylor invented almost everything in one form or another that we use today in the office and at home." Taylor led Xerox PARC during its formative period. For Taylor, collaboration was critical to the success of his products. Taylor and the rest of his team at PARC garnered insights through group creativity, and Taylor often emphasized the group component of his teams' work at Xerox PARC.[62]

---

[61]Kim-Pang, Alex Soojung (2000). "The Xerox PARC visit." Making the Macintosh: Technology and Culture in Silicon Valley. Created: July 14, 2000. Accessed August 28, 2019. https://web.stanford.edu/dept/SUL/sites/mac/parc.html.

[62]Berlin, Leslie. "You've Never Heard of Tech Legend Bob Taylor, but he Invented Almost Everything." Wired. Last updated April 21, 2017. Accessed June 7, 2019. www.wired.com/2017/04/youve-never-heard-tech-legend-bob-taylor-invented-almost-everything/.

While Lick and Taylor poured the foundation, a group of scientists built on that, recognizing that there was an applied psychology to humans interacting with computers. A "user-centered" framework began to emerge at Stanford and PARC; this framework was eventually articulated in the 1983 book *The Psychology of Human-Computer Interaction*[63] by Stuart Card, Thomas Moran, and Allen Newell. Though the book predates the widespread presence of personal computing—let alone the Internet—it tightly described human behavior in the context of interacting with a computer system.

The expression of core concepts such as showed that the computer as an interlocutor was now in play and restates Lick's vision of 1960.

> *The user is not an operator. He does not operate the computer; he communicates with it to accomplish a task. Thus, we are creating a new arena of human action: communication with machines rather than operation of machines.* (Emphasis theirs)[64]

The Psychology of Human-Computer Interaction argued that psychological principles should be used in the design phase of computer software and hardware in order to make them more compatible with the skills, knowledge, capabilities, and biases of their users.[65] While computers are, ultimately, tools for humans to use, they also need to be designed in a way that would enable users to effectively work with them. In short, the fundamental idea that we have to understand how people are wired and then adapt the machine (i.e., the computer) to better fit the user arose from Card, Moran, and Newell.

Alan Newell also had a hand in some of the earliest AI systems in existence; he saw the computer as a digital representation of human problem-solving processes.[66] Newell's principal interest was in determining the structure of the human mind, and he felt that structure was best modeled by computer systems. By building computers with complex hardware and software architectures, Newell intended to create an overarching theory of the function of the human brain.

Newell's contributions to computer science were a byproduct of his goal of modeling human cognition. Nevertheless, he is one of the most important progenitors of AI, and he based his developments on psychological principles.

---

[63]Card, Stuart K., Moran, Thomas P., and Newell, Allen. *The Psychology of Human-Computer Interaction* vii–xi, 1–19. Hillsdale, NJ: Lawrence Erlbaum Associates (1983).

[64]Card et al. *Psychology*, 7.

[65]Card et al. *Psychology*, 11–12.

[66]Piccinini, Gualtiero. "Allan Newell." University of Missouri-St. Louis. www.umsl.edu/~piccininig/Newell%205.htm. Accessed June 25, 2019.

PSYCHOLOGY AND COMPUTING SHOULD GO HAND IN HAND

BOB: There's an important dialog between psychologists who were trying to model the mind and brain and computer scientists who were trying to get computers to think.

GAVIN: Sometimes, it seems like the same people were doing both.

BOB: Right—the line between computer scientists and cognitive psychologists was blurred. But you had people like Newell and others who saw the complex architecture of computers as a way to understand the cognitive architecture of the brain.

GAVIN: This is a dance. On one hand, you have computer scientists building complex programs and hardware systems to mimic the brain, and on the other hand, you have psychologists who are trying to argue how to integrate a human into the system.

A simple example is the old "green screen" cathode-ray tube (*CRT*) monitor, where characters lit the screen up in green. One anecdotal story had the hardware technologists pulling their hair out because the psychology researchers argued that the move from all caps font to mixed-case font would be better from a human performance perspective if the screen had black characters on a white background. This was a debate because the hardware technology required to make it easier for the human is vastly different from a CRT. Even with this story, you can imagine how having computer scientists and psychologists in the same room advanced the field.

BOB: It's really the basis of where we're at today. Even though computers and brains don't work the same way, the work of people like Allan Newell created insights on both sides. Especially on the computing side, conceptualizing a computer as being fundamentally like a brain helped make a lot of gains in computing.

GAVIN: Psychology and computer science can work hand in hand.

BOB: Ideally, they would. But it doesn't always happen that way. For instance, in today's world, most companies are hiring computer scientists to do natural language processing and eschewing linguists or psycholinguists. Language is more than a math problem.

The point   Psychology and computing should go hand in hand. In the past, computer scientists with a psychology background generated new, creative insights.

Bouncing back from failure

The AI winter(s) can be understood through a "hype curve" for AI laid out by AI researcher Tim Menzies.[67] Menzies says that AI, like other technologies, has reached a "peak of inflated expectations" early in its career (in the mid-1980s). This was the result of a quick rise to prominence and overoptimism. Once those who had believed AI's hype discovered that it still had a long way to go, this was followed by a "trough of disillusionment" (the AI winter); see Figure 2-1. However, this trough did not last all that long. By 2003, when Menzies was writing, he felt AI had made a slow rise toward a level of success above the trough but below the peak, or a "plateau of profitability."

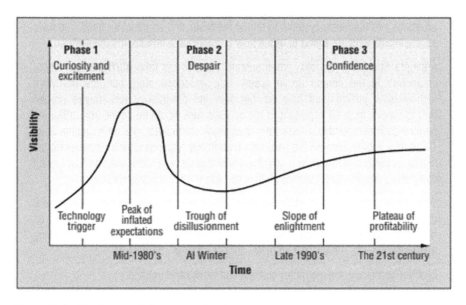

Figure 2-1. The hype cycle for new technology

AI's gradual climb back to solvency after the AI winter of the late 1980s is a story of rebirth that can also provide lessons for us. One of the key components of this rebirth was a type of AI called neural networks that we alluded to earlier. Neural networks actually date back to at least the 1950s,[68] but they became popular in the 1990s, in the wake of the AI winter as a means to continue AI research under a different name.[69] They included an emphasis on

---

[67]Menzies, Tim. "21st-Century AI: Proud, Not Smug." IEEE 3 (2003): 18–24.
[68]https://towardsdatascience.com/a-concise-history-of-neural-networks-2070655d3fec (retrieved July 30, 2019).
[69]Bostrom, *Superintelligence*, 9.

and newfound focus on a property of intelligence that Schank emphasized in 1991. Schank argued that "intelligence entails learning,"[70] implying that true AI needs to be able to learn in order to be intelligent. While expert systems had many valuable capabilities, they were rarely capable of machine learning. **Artificial neural networks** offered more room for machine learning capabilities.

 Definition   **Artificial neural networks**, sometimes simply called neural networks, is a type of AI system that is loosely based on the architecture of the brain, with signals sent between artificial neurons in the system. The system features layers of nodes that receive information, and based on a calculated weighted threshold, information is passed on to the next layer of nodes and so forth.[71]

Neural networks come broadly in two types: supervised and unsupervised. Supervised neural networks are trained on a relevant dataset for which the researchers have already identified correct conclusions. If they are asked to group data, they will do so based on the criteria they have learned from the data on which they were trained. Unsupervised neural networks are given no guidance about how to group data or what correct groupings might look like. If asked to group data, they must generate groupings on their own.

Neural networks also had significant grounding in psychological principles. The work of David Rumelhart exemplifies this relationship. Rumelhart, who worked closely with UX pioneer Don Norman (among many others), was a mathematical psychologist whose work as a professor at the University of California-San Diego was similar to Newell's. Rumelhart focused on modeling human cognition in a computer architecture, and his work was important to the advancement of neural networks—specifically back propagation which enabled machines to "learn" if exposed to many (i.e., thousands) of instances and non-instances of a stimulus and response.[72]

Feigenbaum said that "the AI field...tends to reward individuals for reinventing and renaming concepts and methods which are well explored."[73] Neural networks are certainly intended to solve the same sorts of problems as expert systems: they are targeted at applying the capabilities of computer technologies to qualitative problems, rather than the usual quantitative

---

[70]Schank, 40.

[71]Hardesty, Larry (2017). "Explained: Neural networks. Ballyhooed artificial-intelligence technique known as 'deep learning' revives 70-year-old idea." MIT News. http://news.mit.edu/2017/explained-neural-networks-deep-learning-0414. Posted April 14, 2017. Accessed. August 28, 2019.

[72]Remembering David E. Rumelhart (1942–2011). Association for Psychological Science. Accessed July 8, 2019. www.psychologicalscience.org/observer/david-rumelhart.

[73]Feigenbaum, "Knowledge Engineering," 10.2.

problems. (Humans are very good at qualitative reasoning; computers are not good at all in this space.) Supervised neural networks in particular could be accused of being a renamed version of expert systems, since the training data they rely on could be conceived of as a knowledge base and the neuron-inspired architecture as an inference engine.

There is some truth to the concept that AI's comeback is due to its new name; this helped its re-adoption in the marketplace. After all, once a user (individual, corporate, or government) decides that "expert systems," for example, do not work for them, it is unlikely that they'll want to try anything called "expert systems" for a long time afterward. If we want to reintroduce AI back into their vocabularies, we must have some way of indicating to these users that a technology is different enough from its predecessor to be worth giving another chance. The rise of a new subcategory of AI with a new name seems to have been enough to do that.

However, slapping a new name on a similar technology is not enough to regain users' trust on its own. The technology must be different enough from its predecessor for the renaming to seem apt. Neural networks were not simply a renamed, slightly altered version of expert systems. They have radical differences in both their architecture and their capabilities, especially for those neural networks which allow a neural network to adjust the weights of its artificial neurons according to the effectiveness of those neurons in producing an accurate result (i.e., "back propagation"). This is just the sort of learning Schank believes is essential to AI.

Norman and the rise of UX

As AI morphed, so did user experience. The timing of the rise of neural networks (in the early 1990s) roughly coincides with Don Norman's coining of the term "user experience" in 1993. Where HCI was originally focused heavily on the psychology of cognitive, motor, and perceptual functions, UX is defined at a higher level—the experiences that people have with things in their world, not just computers. HCI seemed too confining for a domain that now included toasters and door handles. Moreover, Norman, among others, championed the role of beauty and emotion and their impact on the user experience. Socio-technical factors also play a big part. So UX casts a broader net over people's interactions with stuff. That's not to say that HCI is/was irrelevant; it was just too limiting for the ways in which we experience our world.

But where is this leading? As of today, UX continues to grow because technologies build on each other and the world is getting increasingly complex.[74] We come into contact daily with things we have no mental model for, interfaces that present unique features, and experiences that are richer and deeper than they've ever been. These new products and services take advantage of new technology, but how do people learn to interact with things that are new to the world? These new interactions with new interfaces can be challenging for adoption.

More and more those interfaces contain AI algorithms. A child growing up a decade from now may find it archaic to have to type into a computer when they learn from the very beginning that an AI-natural-language-understanding Alexa can handle so many (albeit mundane) requests. We may not recognize when our user experience is managed by an AI algorithm. Whenever a designer/developer surfaces the user interface to an AI system, there is an experience to evaluate. The perceived goodness of that interface (the UX) may determine the success of that application.

The point   As UX evolves from HCI, it becomes more relevant to AI.

Ensuring success for AI-embedded products

For those who code, design, manage, market, maintain, fund, or simply have interest in AI, having an understanding about the evolution of the field and the somewhat parallel path of UX is necessary to explore. Before moving on to Chapter 3, which will address some of today's verticals of AI investment, a pause to register an observation is in order: there is a distinct possibility that another AI winter is on the horizon.

The hype is certainly here. There is a tremendous amount of money going into AI. Commercials touting the impressive feats of some new application come at us daily. Whole colleges are devoting resources, faculty, students, and even buildings toward AI. But as described earlier, hype can often be followed by a trough.

In many ways, we need to return to Licklider's perspective that there exists a symbiosis between the human and the machine. Each has something to contribute. Each will be more successful if there is an understanding about who does what, when, and how. For AI to succeed, to avoid another winter, it needs good UX.

---

[74]Nielsen, Jakob (2017). "A 100-Year View of User Experience." Last updated: December 24, 2017. Last accessed July 30, 2019. www.nngroup.com/articles/100-years-ux/.

AI is at an important crossroads. In order to understand how AI can achieve success, we must set one key assumption: Let's assume the underlying code— the algorithms—work. That it is functional and purposeful. Let's assume also that AI can deliver on the promise and opportunity. The challenge is whether success will follow. Know that success does not simply happen; it needs to be carefully developed and properly placed. Google was not the first search engine. Facebook was not the first social media network. There are many factors that make one product a success and another a historical footnote.

The missing element is not the speed of AI or even the possible uncovering of patterns previously unknown. It will be on whether the product you build can take advantage of the insight or idea and be successful. The key position put forth is that AI is here, but much around it needs to be shaped, developed, and made more usable. The moment is here where the two fields developing in parallel should converge.

THE CONFLUENCE OF AI AND UX

BOB: At what point does our story of the computer scientist and psychologist merge? The term HCI itself embodies not just computers but the interaction with humans.

GAVIN: Again, the dance follows the AI timeline. From the beginning, there was convergence on the future of computing and how it may integrate with humanity. Neural networks brought ideas that computer networks may mimic the brain. Programmers build AI systems and cognitive psychologists focused on making technology work for people.

BOB: The time is now. We started our careers over 25 years ago pleading for a budget to make the technology "user friendly" or "usable." Today, the value of a good product experience is not just nice to have, but clearly linked to the brand experience and tied to company value.

GAVIN: People speak about the Apple brand with high reverence. Much time and effort went into the design of not only the product (e.g., iMac, iPod or iPhone), but the design of the brand experience. The Apple brand almost transcends a singular product. Perhaps out of necessity, businesses recognize the value of the brand experience—as AI embeds itself into new products, good design matters. Focus on the experience is where the emphasis needs to be. No one cares about the AI algorithm or if an unsupervised neural net was used. People care about good experiences. Or more aptly, people pay for good experiences.

The point   HCI's time is now. AI technology is at a point where the differentiator for success is the user experience.

Conclusion: Where we're going

In the next chapter, we will explore AI at the 30,000 foot level. This will describe core elements of how AI works so we can explore areas where we can apply UX to improve AI outcomes.

It is important to note that the emphasis will not be detailed technical information; in fact, let's assume the AI algorithms at the heart of it all work just fine. What can we do as product managers, marketers, researchers, UX practitioners, or even technophiles to understand the potential gaps and opportunities for nonprogrammers and non-data scientists to impact product success? The next chapter is about all of the areas where AI is emerging and to identify where a better experience would make a difference.

AI-Enabled Products Are Emerging All Around Us

Technology is everywhere

Access to computing power is at our fingertips. In the palm of our hands, the mobile phones we hold are smarter than desktop machines that sat on our office spaces a decade ago. Still in our lifetime, the small screen that we stare at to read news, check email, or play a game has the computing power that used to take up an entire room as a mainframe computer. With this power comes connectivity that gives us access to information. The convergence of computing power, connectivity, and data opens the doors to so much more. Simplistically, these connected devices form what is called the Internet of Things (IoT). And now, companies are embedding many of these devices with AI. What started with voice-enabled platforms connecting to IoT devices now has broader implications of bringing intelligence to what would be **ubiquitous computing**.

© Gavin Lew, Robert M. Schumacher 2020
G. Lew and R. M. Schumacher, *AI and UX*,
https://doi.org/10.1007/978-1-4842-5775-3_3

▨ **Definition**   **Ubiquitous computing** describes a state in which we interact with computers throughout our daily lives without even considering them as computers.[1]

UBIQUITIOUS COMPUTING

GAVIN: As we enter into the 2020s, we are creeping ever closer to a state of **ubiquitous computing**. The rapid growth of the Internet of Things (IoT) and increased connectivity could mean AI could learn from your behaviors and predict your habits. For example, your thermostat in your house might know you like to keep the temperature at 73 degrees Fahrenheit when you are home. When you leave for work, the system reduces the temperature, but as you drive home, the system turns it back up. *You don't even think twice about the devices or connectivity to make this happen.* It just works. You walk into the house, and *ahhh*.

BOB: You said a crucial part at the end there: "it just works." For it to "just work," the UX has to build trust. The virtual assistant has to understand exactly what your query means, communicate it to your thermostat, and do it in a timely and consistent fashion so that you'll learn to trust it. Only if you trust it will you feel like it "just works."

▨ **The point**   We are at a unique moment where ubiquitous computing is here, but what will make the difference is the *experience*.

Moving toward ubiquitous computing requires us to develop products with the kind of UX that reduces barriers between the user and the device. In the medical technology space, there are already ubiquitous computing tools—medical instruments that are connected to the Internet of Things. These are not necessarily AI in and of themselves, but their data could be fed into AI systems. There are already Internet-enabled medical instruments numbering in the millions,[2] and there will likely only be more in the years to come. The number of things that are "smart" or "Internet enabled" is in the billions. Just walk through the Consumer Electronics Show in Las Vegas every year and you'll see some really far out stuff that's smart: fishing rods, fabrics, forks, and soccer balls to name a few.

---

[1]Witten, Bekah. "Ubiquitous Computing: Bringing Technology to the Human Level." USF Health https://hscweb3.hsc.usf.edu/is/ubiquitous-computing-human-technology/.
[2]Marr, Bernard. "Why the Internet of Medical Things (IoMT) Will Start To Transform Healthcare in 2018." Last updated January 25, 2018. Accessed June 1, 2019. www.forbes.com/sites/bernardmarr/2018/01/25/why-the-internet-of-medical-things-iomt-will-start-to-transform-healthcare-in-2018/#75cf88e54a3c.

AI systems will find their role in the interplay across non-intelligent devices, AI systems (themselves), and human beings. For one thing, AI systems may be powerful enough to collect and synthesize the massive amounts of data thrown off by a large collection of non-intelligent IoT devices. For another, designers can take a lesson from these non-intelligent devices by giving their AI systems a distinct role and use cases for the user. If the user begins to take these use cases for granted, the barriers between the users and the AI may start to fall, which is a good thing.

AI has an important role to play in the matrix of humans and devices that will rise in the next decade. It goes without saying that things are moving quickly—very quickly. While expert systems and neural networks are still underlying some of our most important technology systems, they have been succeeded as buzzwords by other terms: virtual assistants, deep learning, natural language processing/understanding, and more. IBM cites[3] "dynamic intelligence" as a strength of its Watson Health program, a leader in the medical AI space. As Roger Schank predicted, learning and adaptability have become important parts of AI development.

Neural networks continue to provide a viable architecture for AI development, though they have taken on yet another name. "Deep learning" is one of the most common terms we hear coming from the AI side of things today. Deep learning systems are essentially machine learning-capable, multilayered, unsupervised neural networks, according to the hardware company NVIDIA.[4] The major difference between them and the neural networks of the 1990s is that today's deep learning systems are far more capable of "learning" without human intervention than any past systems.

Deep learning systems have been applied across a wide variety of fields, and we don't have time to address them all here. We will address some of the more glamorous subfields of AI today, including virtual assistants and self-driving cars. However, to begin, we want to look at one of the most important areas where AI is making an impact: medicine.

Medical AI as a team player

Healthcare is a complex industry. Because it involves human lives, the stakes are naturally high, but so is the potential for high reward and opportunity. Companies have targeted healthcare as an area to integrate AI products and services.

---

[3]"About IBM Watson Health." Accessed May 29, 2019. www.ibm.com/watson/health/about/.

[4]"Deep Learning." NVIDIA Developer. Accessed May 30, 2019. https://developer.nvidia.com/deep-learning.

AI is real, it's mainstream, it's here and it can change almost everything about healthcare.

—Virginia Rometty, CEO, IBM[5]

HEALTHCARE IS COMPLEX AND AI MIGHT NOT HAVE ALL THE BUILDING BLOCKS

BOB: IBM Watson, an AI system intended for healthcare applications, kicked off a marketing launch that in some ways did too well.

GAVIN: It started when IBM Watson defeated two human contestants in a game of Jeopardy in 2011. Following that victory, IBM announced that AI had mastered natural language and would now take on healthcare, a leap that at face value left a lot more questions than answers.

BOB: But from that moment on, IBM Watson was marketed as an AI doctor. In 2014, IBM did demonstrations where Watson was fed a bag full of symptoms and out came a list of possible diagnoses with confidence levels and links to supporting medical literature. The CEO said this was the start of a new "Golden Age."[6]

GAVIN: Generalizing natural language to play Jeopardy is one thing. Unfortunately, the billions spent since 2011 were heavily focused on being a doctor—to take in symptoms and output diagnosis and treatment plans. Watson set the target to be as good a doctor.

BOB: This might be a case of believing your own press. When you evoke messages of ushering in a revolution in healthcare, the course you set might be a bit too ambitious.

The point   Healthcare is complex and AI may still not be up to the standard required.

IBM Watson was discussed in Chapter 1 in the oncology example where Watson predicted cancer treatment recommendations well for US doctors, but not for South Korean doctors. This was an example of how two separate AI outcomes were compared instead of merging the data to simply set

---

[5]Strickland, Eliza (2019). "How IBM Watson Overpromised and Underdelivered on AI Health Care." IEEE Spectrum. Last updated April 2, 2019. Last accessed November 6, 2019.https://spectrum.ieee.org/biomedical/diagnostics/how-ibm-watson-overpromised-and-underdelivered-on-ai-health-care.

[6]Strickland, Eliza. (2019). "How IBM Watson Overpromised and Underdelivered on AI Health Care." IEEE Spectrum. Last modified April 2, 2019. Accessed August 19, 2019. https://spectrum.ieee.org/biomedical/diagnostics/how-ibm-watson-overpromised-and-underdelivered-on-ai-health-care.

geographic **context** (i.e., flagging data that were US cases and those that were South Korean). Watson was admonished for not correlating to recommended South Korean treatment plans. We argued that AI found insight—that something different is happening between the US and South Korean oncologists. Perhaps identifying this finding could lead to better health outcomes. Healthcare is complex and there is an opportunity for AI to uncover insights using data already collected.

Now, let's look at medical AI from a different angle. We think medical AI provides an effective illustration of how AI can overcome fear and disillusionment. Specifically, medical AI deals with one of the most important fears surrounding AI's proliferation in general: the idea that people will lose their jobs to automated systems.

At its inception in 2011, a spokesperson for the company said IBM Watson would make diagnoses based on some 200 million pages of data, including academic research, insurance claims, and medical records.[7] If this had come to pass, medical AI would have been capable of doing many of the things that doctors do on a day-to-day basis (though perhaps not with the same level of accuracy, as we'll discuss).

Naturally, this allowed onlookers to envision a day where some doctors—not to mention other medical professionals—are put out of a job. The doctors at highest risk seemed to be those whose work is more focused on the sorts of analytical tasks that AI does well (such as radiologists, who are experts in analyzing medical scans). In 2016, the University of Pennsylvania physician Saurabh Jha predicted that radiologists would lose their jobs to AI within "10 to 40 years, but closer to 10 years."[8]

As it turns out, IBM Watson in particular was not as capable as advertised—staving off any fears of AI-induced consequences for employment in medicine. Daniela Hernandez and Ted Greenwald authored a concerning article in *The Wall Street Journal* that we described in Chapter 1.[9] Hernandez and Greenwald cited difficulties with inconsistent data formatting and the still-developing state of cancer research as possible reasons why IBM Watson has stagnated. IBM Watson, in its current state, only accesses research, not medical records or insurance claims. The dream of unsupervised or lightly supervised neural networks diagnosing cancer on their own still awaits access to patient information and patient outcomes.

---

[7]Mearian, "IBM's Watson…to diagnose patients."

[8]Jha, Saurabh. "Will Computers Replace Radiologists?" Last updated May 12, 2016. Accessed July 14, 2019. www.medscape.com/viewarticle/863127#vp_3.

[9]Hernandez, Daniela and Ted Greenwald. "IBM has a Watson dilemma." The Wall Street Journal. www.wsj.com/articles/ibm-bet-billions-that-watson-could-improve-cancer-treatment-it-hasnt-worked-1533961147.

In 2018, it became clear that IBM had overhyped its own product in the initial stages.[10] Its developers may have thought it would develop more quickly than it actually did. Specifically, IBM Watson's oncology program is reportedly too often inaccurate to be reliable. One healthcare industry expert interviewed by Mearian said that IBM Watson was released to the world too early. It still needed more time to develop its knowledge base.[11] IBM disputes the claims that Watson is inaccurate, saying that it has helped a significant portion (2–10%) of cancer patients to change to a different treatment and that it often agrees with physician recommendations.[12]

New York Medical College's Douglas Miller and IBM's Eric Brown co-authored a study in *The American Journal of Medicine* that assuaged fears about AI taking medical jobs.[13] Though Miller and Brown did not rule out the possibility of unforeseen consequences arising in the future of AI, they cited issues with accuracy and intuition as reasons that AI is not yet set to overtake any physicians. Rather, Miller and Brown recommended that physicians use medical AI as a powerful "tool" which can assist them in diagnosis and treatment.[14]

The point   At this juncture, AI augments human cognition in medicine (and most other areas) but does not replace it.

If medical AI can follow Miller and Brown's advice, it can avoid falling into a hype-driven freefall like the one that machine translation experienced in the late 1960s. In the past decade, medical AI certainly overpromised and underdelivered. But it is still delivering something that can be useful to doctors. The question is whether developers and doctors can adapt to the new world and build a product that is more appropriate considering AI's current capabilities.

As it turns out, doctors are already doing their part. A Department of Veterans Affairs doctor cited by Hernandez and Greenwald said the service can be helpful in the search for relevant academic research: "Dr. Kelley said Watson's recommendations can be wrong, even for tried-and-true treatments. On the other hand, he said, it is fast and useful at finding relevant medical articles, saving time and sometimes surfacing information doctors aren't aware of."[15]

---

[10]Mearian, Lucas. "Did IBM overhype Watson Health's promise?" Computerworld. Last updated November 14, 2018. Accessed May 31, 2019. www.computerworld.com/article/3321138/did-ibm-put-too-much-stock-in-watson-health-too-soon.html.

[11]Mearian, "Did IBM overhype…"

[12]Hernandez and Greenwald.

[13]Miller, D. Douglas, and Eric W. Brown. "Artificial Intelligence in Medical Practice: The Question to the Answer?" The American Journal of Medicine, 131/2(2018): 129–133. https://doi.org/10.1016/j.amjmed.2017.10.035.

[14]Miller and Brown, "Artificial Intelligence in Medical," 132.

[15]Hernandez and Greenwald.

Medical AI can focus on what it does well—analyzing a large corpus of research for relevant articles, for example—allowing doctors to spend more time working on the diagnostic tasks that AI isn't capable of yet. It can also serve as a second-opinion generator. An AI diagnosis shouldn't be the only diagnosis a patient receives, but it can supplement a doctor and perhaps point out if there is something else to consider. This could be a vitally important tool in the United States, where a high percentage of patients with serious conditions are often misdiagnosed.[16] Medical AI will present an external voice that can help reduce error frequency.

TRAINING AI TO LEARN IS JUST THE FIRST STEP

GAVIN: Interestingly, IBM Watson and its goal of consuming mass volumes of medical literature to look for patterns may have been its first misstep.

BOB: Scanning articles for patterns is by definition a form of AI, but how does it map to how a doctor reads articles? Does AI value the same parts of a paper that a human doctor values? Do the data scientist and AI developer believe that the truth will wash out and true knowledge can be uncovered, or is the practice of medicine more nuanced?

GAVIN: AI looks for correlations and statistical patterns. So, decades of medical literature would set precedent. But, what about the latest study using a new therapy, like gene therapy? What if this new direction changes everything? Eventually, scores of peer-reviewed publications will be written, but today, for patients suffering, how much weight would AI place on the latest groundbreaking study? How does AI differentiate groundbreaking from an isolated study amongst volumes of peer-reviewed and non-peer reviewed articles? AI in healthcare must be in the business of improving health outcomes for patients today.

The point   Just training AI to learn using both historical research data and cutting-edge novel treatments recently published. This is just the first step in the complex walk that is healthcare. Doctors need to **trust** AI and then incorporate its insights into practice.

In medicine, AI would be better served if it assisted human beings, rather than project the goal of replacing them. Lick, discussed in Chapter 2, would have loved this application of his AI principle. Medical AI has the potential to increase doctors' accuracy, while not threatening their job security. Many of the products in the medical AI space are designed specifically with this purpose in mind. Eric Topol, the author of a book on medical AI, envisions a future in which AI helps

---

[16]20 percent of patients with serious conditions are first misdiagnosed, study says. www.washingtonpost.com/national/health-science/20-percent-of-patients-with-serious-conditions-are-first-misdiagnosed-study-says/2017/04/03/e386982a-189f-11e7-9887-1a5314b56a08_story.html. Bernstein, Lenny. April 4, 2017 (retrieved May 22, 2020).

doctors to spend more time working with individual patients, rather than spending their days reading dozens of scans, without the time to focus on any particular patient.[17] Topol sees doctors offloading their routine tasks and having more time for the sort of "intuitive" work that Miller and Brown argue humans are best suited to do. He cites AI programs that are already working to diagnose specific diseases, such as an algorithm that detects diabetic retinopathy, a vision-impairing condition caused by diabetes. Topol's vision of AI would lead to a collaborative relationship in which medical professionals and medical AI each take on the portion of their work that they are best suited to do.

AI AS JUST PART OF A TEAM

GAVIN: If AI works together with human beings, they can do more as a team than either one could do individually. That should be the model for AI everywhere.

BOB: If AI is seen as collaborating with a human user, maybe it won't seem so scary or dangerous. It'll seem more helpful.

GAVIN: In the medical AI space, it seems like we backed into that relationship. Initially, there was a lot of hype surrounding medical AI, and it was supposed to do all these amazing things. But that turned out to be easier said than done.

BOB: It reminds me of machine translation in the 1950s and 1960s. The computer scientists of the 1960s thought that once they had taught a machine to translate a few Russian phrases into English, teaching it to become a fluent translator would come along quickly. The designers of medical AI might have made the same mistake. It's possible to get AI to do certain tasks in the medical space, but making it a generalized medical intelligence is way more difficult.

GAVIN: But it seems like there's a path for medical AI to avoid the domain-specific AI winter that plagued machine translation after the ALPAC report. If it can concentrate on what it does well—for now, combing through its archives for medical research or offering a second opinion—it can be very useful.

BOB: In some sense, the "irrational exuberance" clouded reality and led many to underestimate the difficulty of the problems. Because the systems are still developing, there has to be an allocation of responsibility. There should be some ability for doctors to say, for example, "For this type of cancer, I don't want a treatment plan right now, can you just show me the relevant research?" But maybe for this other type of cancer, the treatment plans are useful. And then maybe the AI can learn what kind of feedback to give its users.

GAVIN: This collaboration between AI and doctor could result in better health outcomes. Imagine, a doctor asking for only the newest treatments in a rare condition that is

---

[17]Belluz, Julia. "3 ways AI is already changing medicine." Vox. Last updated March 15, 2019. Accessed May 31, 2019. www.vox.com/science-and-health/2019/3/15/18264314/ ai-artificial-intelligence-deep-medicine-health-care.

spreading quickly. But, this is where the design matters. Programmers can develop the next wave of medical AI with interactions in mind.

 The point   AI can be dangerously overhyped as being more capable than a human in a complex job. But AI's real value is as a helpful member of the team if it can adapt to its strengths and limitations.

Healthcare is a good example of where buying into the technology can be too appealing. Sometimes we build applications based on poor assumptions, even a hunch, and not based on true need. Similar to how holding a hammer makes every problem looks like a nail. But in AI's case, it is, "Here is technology and let's apply it everywhere," or worse, the dreaded, "if you build it, they will come" fallacy. Martin Kohn, who was the chief medical scientist at IBM during the Jeopardy period, was once so enamored by the technology's potential that he called it a trap. "Merely proving that you have powerful technology is not sufficient," he says. "Prove to me that it will actually do something useful— that it will make my life better, and my patients' lives better."[18] Since Kohn has left IBM, he still is in search of peer-reviewed papers in the medical journals to demonstrate that AI can improve patient outcomes and save health systems money. "To date there's very little in the way of such publications," he says, "and none of consequence for Watson."[19]

The rise of virtual assistants

In Chapter 2, we discussed the initial stagnation of **voice and virtual assistants**. After Siri was released in beta form, the assistant's limited capabilities led to public frustration with the service and with virtual assistants in general. This led to what we'd consider a domain-specific **AI Winter**—one that had a dramatic effect on Microsoft's Cortana in particular. Yet the emergence of Amazon's Alexa opened the door to try voice again.

Let's explore what it took to restore the impression of what a **voice assistant** could and could not do. What considerations in research and design should we attribute to Alexa's success?

---

[18]Strickland, Eliza (2019). "How IBM Watson Overpromised and Underdelivered on AI Health Care." IEEE Spectrum. Last updated April 2, 2019. Last accessed November 6, 2019. https://spectrum.ieee.org/biomedical/diagnostics/how-ibm-watson-overpromised-and-underdelivered-on-ai-health-care.

[19]Milanesi, Carolina (2016). "Voice Assistant, Anyone? Yes please, but not in public!" Creative Strategies. Last modified June 3, 2016. Accessed August 23, 2019. https://creativestrategies.com/voice-assistant-anyone-yes-please-but-not-in-public/.

BEYOND SIRI

GAVIN: Let's talk about Siri, our favorite virtual assistant that most know about but rarely use.

BOB: I would argue that more people accidentally wake Siri than those who proactively engage it to actually solve tasks.

GAVIN: That's the problem. And I have spoken to designers of other voice-based virtual assistants, and they go on about how different their system is from Siri. They talk about how Siri treats errors like a joke. But, instead, their system is unique because it took a different approach, one that presented a more *mature* interaction. Some spoke of *triggered interactions* where after so many successful instances of the same voice command, hints would be added to responses to reveal new features or shortcuts. After a failed voice command where some information was recognized, the voice response would change to include help. While this may or may not improve **interactions**, the problem was the lack of adoption that really hurt potentially novel features from evolving further.

BOB: Imagine the millions of dollars that went into developing Microsoft's Cortana or Samsung's Bixby. How many users never even tried to use them? And it probably wasn't solely because they didn't like Microsoft or Samsung, but maybe they assumed the experience would be just like using Siri?

GAVIN: This is what people do when they experience frustration or cannot get a product to work. People often generalize the experience to other similar products. This is a domain-specific **AI winter**. Novel approaches to voice assistants do not advance simply because they never had the chance.

BOB: At least in the case of Amazon Fire's failure, sometimes smart ideas can find new life in a different form factor--one that sat like an obelisk on the kitchen table.

GAVIN: Amazon Echo breathed new life into voice assistants because a C-level executive like Jeff Bezos took the risk when he saw the opportunity.[20]

The point Designing a winning product depends on many factors, the least of which involves seizing opportunities to advance novel features. Look at older product designs to find features that deserve a second chance.

---

[20]Bariso, Justin (2019). "Jeff Bezos Gave an Amazon Employee Extraordinary Advice After His Epic Fail. It's a Lesson in Emotional Intelligence. The story of how Amazon turned a spectacular failure into something brilliant." Inc. Last updated December 9, 2019. Accessed May 14, 2020. www.inc.com/justin-bariso/jeff-bezos-gave-an-amazon-employee-extraordinary-advice-after-his-epic-fail-its-a-lesson-in-emotional-intelligence.html.

The revolution in development and deployment of virtual assistants came from the confluence of (1) vastly improved natural language processing and natural language understanding, (2) high speed Internet, (3) cloud computing, and (4) ubiquitous deployment of tiny microphones and speakers. But even these things did not help Siri succeed. What made a material difference for Alexa was that it was embedded in a standalone device: the Echo. Mobile virtual assistants differ, because they have other primary uses, and the virtual assistant function is just a new feature within a larger ecosystem.

This means that the Echo, unlike the iPhone or PC, was designed for the specific purpose of serving as an effective virtual assistant. The Echo was designed with context of use in mind. Its cylindrical form, combined with marketing, meant that the Echo had clearly defined use cases in the home.

There's a clear lesson to be found in the Echo's success: a product with clearly defined use cases was effective in helping consumers understand the context of use for a new product category. Let's explore three contexts a little deeper.

Context of use

Using a virtual assistant in the home is no big deal. Using it at a ballpark is a no-go. For Alexa on the Amazon Echo platform, the task of incorporating context of use is easy: the technology simply has to be designed to incorporate a single-use category. In fact, the Echo will likely stay in the same room for the duration of its use. (Not to mention that Amazon can sell a lot more if you need one in each room of the house!) Perhaps it could adapt based on whether it is in the bedroom or the kitchen. But for an assistant that primarily lives on the phone, like Siri, the task is more complicated.[21] Siri could be used in the kitchen or the bedroom, but also in the car, in a private office space, or even in some more public setting (a lobby, a public space, etc.) With location services, perhaps it could tailor its responses for the exact location/room where it is being used, but the use case is complicated, given the wide variety of possibilities.

Another part of context of use incorporates not just where the user is, but what they are doing. After all, an Amazon Echo user might be in the kitchen cooking—in which case, a timer or a recipe might be useful for Alexa to have on hand. (Alexa has 60,000 recipes on hand for that situation.)[22] But they also might be in the kitchen chatting with their spouse—in which case, access to their shared calendar might be most important. For assistants on mobile platforms, of course, the contextual possibilities are exponentially higher.

---

[21]Most (all?) virtual assistants can be used on mobile platforms, like phones, as well.

[22]Vincent, James. "Amazon's Alexa can now talk you through 60,000 recipes." The Verge. Last updated November 21, 2016. Accessed July 1, 2019. www.theverge.com/2016/ 11/21/13696992/alexa-echo-recipe-skill-allrecipes.

Even if a virtual assistant is designed to be used in the kitchen, there is still a wide variety of things that a user might do within that context that the assistant needs to be prepared for.

The point   Where you are matters. Allowing AI to know where the user is can improve the experience by making it more accurate and relevant. Designing AI with **context of use** can support AI to be more insightful.

Conversational context

When you're talking to a virtual assistant, it should remember what you were talking about. This is easier said than done. When Business Insider was testing the top four virtual assistants (Siri, Alexa, Google Assistant, and Cortana) back in 2016, they tried asking each virtual assistant about the next Boston Celtics basketball game. All four did fine with that query, but when they asked the follow-up, "Who is their top scorer?", everyone got lost. The assistants didn't understand that "their" was a reference to the Celtics from the previous question. That's an example of assistants missing out on conversational context.[23] And it's not an anomaly in using virtual assistants. Assistants often break the natural flow of conversation by losing track of conversational context—that is, what was just discussed and what pronouns in the current query might refer to that prior information.

The point   We already described **machine translation** and its challenges with language, but designing with an understanding of a temporal context can improve conversations. AI needs to move beyond simple query to task **interactions** and anticipate most likely follow-up queries instead of assuming the conversation only had one question with no follow-up.

Conversational context errors are some of the most frustrating ones that still plague virtual assistants today. It's a difficult problem for humans in normal conversation, let alone for voice dialog designers and programmers. Sometimes, a subsequent query refers to the previous query, and other times, it's an entirely new question. But it's also immensely frustrating to users to ask what seems to be a natural follow-up question and get an unnatural answer.

---

[23]"We put Siri, Alexa, Google Assistant, and Cortana through a marathon of tests to see who's winning the virtual assistant race—here's what we found." Business Insider. Last updated November 4, 2016. Accessed July 1, 2019. www.businessinsider.com/siri-vs-google-assistant-cortana-alexa-2016-11#the-setup-theres-no-perfect-way-to-evaluate-a-talking-ai-database-let-alone-four-of-them-but-i-tried-to-cover-as-many-fundamental-topics-as-i-could-1.

When dialog with an assistant goes bad, it's because the normal conventions of human conversation are not followed. Tracking pronouns is just one instance. There are many others. In fact, most human conversation holds true to certain "maxims"[24] laid out by linguist and philosopher Paul Grice: quantity, quality, relation, and manner. See Table 3-1.

Table 3-1. Grice's four maxims of communication

• Quantity: Information
  o Make your contribution as informative as is required for the current purposes of the exchange.
  o Do not make your contribution more informative than is required.
	• **Quality**: Truth
	  o Do not say what you believe to be false.
	  o Do not say that for which you lack adequate evidence.

• Relation: Relevance
  o Be relevant.
	• Manner: Clarify
	  o Avoid obscurity of expression.
	  o Avoid ambiguity.
	  o Be brief (avoid unnecessary prolixity).
	  o Be orderly.

Voice assistants need to conform to human conversational norms in order to be accepted by human beings. They can do this by adapting to Grice's conversational maxims. To be successful, AI researchers need to embrace what linguists and psycholinguists already know about how people communicate effectively.

APPLYING GRICE'S MAXIMS TO AN ALEXA SKILL

GAVIN: When we say that "AI should be designed to _____," how does this occur? Are we describing a future state or can this be done today?

BOB: Anybody with basic technical skills can try to build a conversation that Alexa can respond to. In fact, I built one in 45 min with no knowledge of how to program an Alexa or experience with the Alexa Skills Kit. Amazon essentially hid the code behind a web interface that allows for skill creation.

GAVIN: So, the barrier of being a developer was removed? You literally dragged and dropped what you wanted Alexa to listen for?

---

[24]Grice, H. P. (1975). "Logic and Conversation," Syntax and Semantics, vol.3 edited by P. Cole and J. Morgan, Academic Press, and Grice, H. P. (1989). Studies in the Way of Words. Harvard University Press.

BOB: Yes. I did not have technical coding skill, but Amazon built an interface to design what Alexa should *say* and what to *listen for* and then what Alexa should *say in response*. So, what we are recommending is to incorporate UX principles in the design.

GAVIN: So, you are not talking about algorithms. Your suggestion is to design a **voice assistant** that follows conversational norms, the kind of things in Gricean maxims.

BOB: This is where UX and AI can merge and make the product much *smarter* because it pulls in UX principles into the design. In this example, **conversational context** would give AI a distinctive edge.

The point   We can learn a lot from what linguists already know about communication patterns. Using Grice's maxims can inform the dialog and lend to anticipated follow-on questions. Analyzing errors can suggest remedy or conversational corrections. These are examples where **conversational context** can be integrated into AI-enabled products.

Informational/user context

This wide-ranging third category of context includes the resources to which a service has access: both resources about the user's attributes and ones that help the user when looking things up. Google, unsurprisingly, is exemplary in both of these categories. They are best at gathering data about you—it's practically their *modus operandi*—and they leverage that data to personalize your results. Similarly, Google has access to the wealth of information that is the first place that most of us go when we want to learn about something: Google. That gives it a distinct advantage over its competitors in terms of access to external information. Alexa and Siri have to rely on outside sources for information about, say, that actor whose most recent movie role you can't quite remember.

Also consider the benefit AI would have knowing the speaker. Alexa and Google Home are moving toward identifying the speaker in the home because this **informational context** of the user is the mom or the dad would presumably play a role in determining the AI-enabled output. For example, the response to the utterance "play some music" would be more relevant if the **voice assistant** knew who was asking.

The point   In simplest terms, **virtual assistants** perform tasks which at the core requires an input (an utterance) and a resulting output (a response). If AI was designed to apply **AI-UX principles** such as **context,** AI could produce much more useful experiences.

Examples: AI-enabled vehicles

All of these types of context are relevant as **virtual assistants** expand to new domains. One space where **virtual assistants** will grow is in mobility (e.g., cars). Currently, cars are one of the most popular use cases for voice-based assistants. According to industry research by Voicebot.ai,[25] **virtual assistants** have about 90 million monthly users on smartphones, about 77 million in cars, and about 46 million on smart speakers.

The car is an interesting use case for the **virtual assistant**. As Vox journalist Rani Molla points out, drivers should not be using touchscreens when driving—and voice assistants seem like a perfect replacement.[26] Today's car voice assistants are often connected through smartphones, which may take advantage of cars with built-in connectivity through programs like Apple CarPlay and Android Auto, but can also function without those built-in features. This gives Siri and Google Assistant an advantage in the auto space, but Amazon is fighting back, planning a plug-in called Echo Auto for cars without built-in voice assistants.[27] As Amazon already has a reputation for making standalone virtual assistants that are built for a particular environment, this may prove to be an effective product.

A few years ago, Toyota previewed a concept car with a more functional automotive virtual assistant called Yui.[28] Yui would have all kinds of unique **informational context** that it would use when helping direct you where to go. If you always visit the grocery store on Tuesdays after work, it would direct you there at the usual time—unless, of course, your digital calendar was booked. It could switch back and forth between allowing the user to drive and driving itself. It would remember your preferences—perhaps remembering whether you tend to prefer the surface streets or the expressway—and use it when recommending places to go and ways to get there. It might even learn those preferences based on its analysis of your facial expressions. That is, it would incorporate a version of **informational context** based on the user.

[25]Molla, Rani. "The future of smart assistants like Alexa and Siri isn't just in homes—it's in cars." Vox. Last updated January 27, 2019. Accessed June 26, 2019. www.vox.com/2019/1/15/18182465/voice-assistant-alexa-siri-home-car-future.

[26]Molla, Rani, "The future."

[27]www.consumerreports.org/automotive-technology/amazon-alexa-isnt-so-simple-in-a-car/.

[28]Etherington, Darrell. "Here's what it's like to drive with Toyota's Yui AI in-car assistant." TechCrunch. Last updated January 6, 2017. Accessed July 2, 2019. https://techcrunch.com/2017/01/06/heres-what-its-like-to-drive-with-toyotas-yui-ai-in-car-assistant/.

UNDERSTANDING CONTEXT IS ESSENTIAL TO SUPPORTING THE USER

GAVIN: Yui is just a concept, and it's a long way from being a reality. But it gives us a preview of what virtual assistants might do in the long term. It's proactive and incorporates different types of context. For context of use, it seems focused specifically on helping the user drive. For informational context, it seems to have a wealth of information about the user and because it is context driven, assistance can be given without the user explicitly requesting a command.

BOB: We've already got the building blocks of proactivity in cars. Think about a feature in many new cars, "lane assist." I recently drove a rental car with this feature. If I drifted outside the current lane without signaling, the car resisted (gently) and moved back to the lane and also played an audio signal. This is a first step in proactivity on the part of the car. It was perhaps more aware of my situation than I was—maybe I was distracted and on my phone, or I was nodding off. This combination of machine vision, sensors, and signaling saves lives and is an example of a good user experience.

GAVIN: Those are the kinds of UX changes we can make in mobility AI in the near term.

BOB: We're still a long way away from developing anything like Yui. Probably 10 years, at least. These are the obvious use cases. I want the AI to either drive my car or be my assistant helping me be aware. For instance, the AI should detect an obstacle in the road even if I don't see it. Or it might even know if the driver next to me is driving like they have had too much to drink; it should help me with my "situational awareness." Let's have AI work to help me drive safely.

GAVIN: But there are UX solutions for virtual assistants in the car that we could adopt today. The car is an easy scenario for context of use. What are users going to want to ask their virtual assistant in the car? Mainly for directions and to play music and podcasts. They can ask whether they've got any new messages. That's really about it.

BOB: True, but that's short-term thinking. Applying what we know about context—large multi-ton vehicles moving at high rates of speed—means that other things should take precedence. If I ask for the weather forecast, should the AI forego detecting an imminent accident and tell me that it's going to rain instead? Of course not.

GAVIN: Well, would that be the same AI system doing both those tasks? Maybe there'd be one AI pumping the brakes when you're at risk of an accident and a whole separate AI that answers your questions.

BOB: I don't know, but if that was the case, they'd certainly have to talk to each other. If I keep getting into close calls on the road while talking to my assistant, the accident-assist AI should be able to tell the AI assistant that it should probably stop letting me ask frivolous questions.

GAVIN: Indeed. Virtual assistants should support the users and understand the priorities of safety (in this case) before information and before entertainment.

BOB: Changes like that would go a long way toward making virtual assistants more useful.

The point   For virtual assistants to be successful, they need to incorporate three types of context—context of use, conversational context, and informational context.

Data science and imputation

Data science, like AI, has been getting a lot of buzz lately.

Definition   Data science—you might also call it data analytics—is the analysis of large datasets to gather insights.

The potential to uncover insights from a massive amount of data that would be impossible by a human is alluring. AI a lot of the time is spent on the algorithms, but, surprisingly, not a lot of time is spent on data. All too often, data is purchased or obtained from historical archives and the data is spun into AI. There is an eagerness to see what happens when the AI algorithm thinks. So much interest is focused on the algorithm and what it uncovers, but are we spending enough attention on the data that are fed into the machine?

AI has been described as a black box; data goes in and what comes out the other side may show no traceability to the inputs (Figure 3-1).

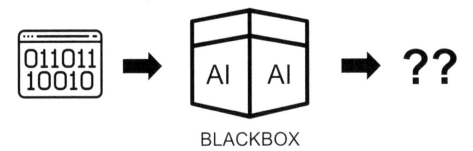

Figure 3-1. Illustration of information entering the black box (i.e., AI process) and insight emerges on the other side

Don Norman, part of the team that developed theories of information processing that led to early neural network development, said this about AI, "The problem with modern artificial intelligence is it's all based on pattern recognition on huge data. It looks for patterns… it reads all the literature, but it doesn't know how to reason about it… There's no understanding."[29] We are blind to what is happening with the algorithms. The patterns formed are merely statistical coefficients. It is impossible to determine the rationale that was used to build the output.

THE SURVEY DATA MINEFIELD

BOB: When our UX consultancy was purchased by one of the top market research firms, we learned firsthand what big data was really about. There was mountains of data—lots of it going back decades. It was all part of being a world-class data company.

GAVIN: Access to 30 years of trended respondent data at that scale must be a beautiful sight for a team building an AI app that could use this data.

BOB: It is actually really useful in understanding trends in market research—and potentially valuable as AI training data. Some of the questions survived multiple changes in methodology from in-home interviews to telephone surveys to online.

GAVIN: And if they were using roughly the same questions, that could be really interesting as an AI data source.

BOB: Well, there is a concern. The people using the output generated from the AI need to know the dataset. How many developers will stop to ask questions about the multiple methods used to collect that data? The data scientists knew how to weigh the impact of differing methods of data collection. But is this lost to AI or just part of the black box?

GAVIN: The dataset has more than data. It has metadata—data about the data. Some might be important but not really factor into the data important to the algorithm and stripped to provide a "clean dataset."

BOB: That's my fear. In any survey, there's almost always data missing whether by design, technical hiccup, or respondent laziness. What data scientists frequently do is fill the missing data using various techniques. This results in a "clean dataset" without missing cells.

GAVIN: Why would data be missing "by design"?

BOB: Back in the 1970s, a 60-minute survey of consumer preferences may have been an interview over coffee in someone's home. In the 1980s, this same survey became a telephone interview, then an online computer survey, and now a survey via mobile phone.

GAVIN: But, who would take a 60-minute survey on a mobile phone?

---

[29]Norman, Don (2016). "Doing design with Don Norman." Medium Podcast. August 24, 2016. Accessed March 18, 2020. https://medium.com/@uxpodcast/design-doing-with-don-norman-6434b022831b.

BOB: That is the point. Because researchers know that few survey takers will spend more than 10 minutes doing a survey on a mobile phone, they break up the survey into sections. Now, it takes several participants to complete what used to happen in the home.

GAVIN: They don't just merge the data. The holes are filled by some equations performed by data scientists. So, you have missing fields being filled by data scientists. Furthermore, some survey methodologists tell me that in some surveys up to 25% of surveys are taken by bots written by people to make "beer money."

BOB: Yup, some datasets have a lot of missing data and others are filled—not by people—but by bots. This should make everyone pause. At its core, AI looks for patterns. Won't AI find the highest pattern match when it correlates data from the underlying algorithms used to fill empty cells or by bots to take surveys? If true, then AI would weigh data made by algorithms over human data?!?!

GAVIN: This is garbage in, garbage out, but we would never know because we cannot peek inside the AI black box to see how, why, or where the output was derived.

The point   This is the problem with AI as a black box. It only is as good as the data that it receives. The data used is critical because if we understand what is going in and then what is happening inside, then the outcome might be just a recreation of what some data scientist used to fill the holes.

We need to be extremely careful when AI uses survey data that is based on human respondents. We actually should approach the data with a high degree of scrutiny when we use the data to train AI algorithms. A marketing firm might build a database of potential customers, or a political campaign might have an internal list of potential voters with lots of parameters and values. This is the data that an AI tool might use to learn from. The data scientist needs to understand how data elements were obtained and if missing data was filled. Knowing the details is vital to the future success of AI because once AI is trained, we cannot look into the "black box" to understand the why or rationale behind the solution. Without accurate data, any AI program that depends on data science's findings is suspicious at best.

Let's review for emphasis this problem: the data scientists often have missing pieces of information in their datasets. When the datasets contain information from surveys or behaviors of human respondents, that raises several issues. Proper analyses should demand that all the cells in the dataset are filled. If researchers simply delete rows of data that have empty cells, their results will come out distorted or biased, especially if there was some confounding variable that the rows with empty cells had in common.[30] So the data scientists need to fill these cells in; this is **imputation**.

■ **Definition**   Imputation is the insertion of values into a dataset, in place of missing cells. Imputation is often needed to run a data analysis.

What's more, survey design often includes built-in **imputation**. Some surveys are so long that it might take 30 minutes to complete. Today, most people are not willing to take a 30-minute survey (and even if they do, the quality of their responses is likely to deteriorate toward the end of the survey. For this reason, survey designers may break up the survey into smaller, say, 10-minute chunks and assign each chunk to an individual subject. Then, using the common data (e.g., demographic information), the researchers would impute these subjects' responses to the chunks of the survey that they did not see. This imputation would be based on the responses to these survey chunks from other subjects who had similar demographic information to the original subject.

Imputation is a necessary, reasonable statistical process for survey research—but could raise havoc as an AI training set. Many large-scale surveys involve at least some sort of imputation. There are a variety of methods for imputing data, many of which include input from an algorithm. These algorithmically generated imputations concern us.

Many data insights are based on trends found in the dataset. If there is an algorithm imputing data, that algorithm will follow a certain pattern in its imputation. This would lead to data which would represent a pattern or trend artificially present in the data. An AI learning system that was fed such a dataset might identify these algorithmic patterns and mistake them for human-generated trends.

This, of course, has potentially dangerous consequences if the algorithmically imputed data is not clearly marked as different in the data analysis.

The solution we see is to mark imputed data as imputed and not strip out important metadata. This will give data scientists context to construct proper training datasets to give to AI developers so AI systems will analyze data as intended. Being skeptical about data gives the AI a better chance to discover something real rather than delivering some artifact inherent in the data.

---

[30]Gelman, Andrew and Hill, Jennifer. "Missing-data imputation." From Data Analysis Using Regression and Multilevel/Hierarchical Models, Cambridge University Press (2006). p. 529–544. Accessed July 3, 2019. doi: https://doi.org/10.1017/CBO97805117 90942.031.

The point   Survey data can often at least partially be generated by algorithms via imputation. Before feeding a dataset based on surveys to AI as training data look into how much of that data is actually generated by an algorithm.

Recommendation engines

If you've ever listened to Spotify's "Discover Weekly" playlist, or Apple Music's similar "New Music Mix," you've interacted with a **recommendation engine**.

Definition   **Recommendation engines** are algorithms that analyze a user's past behavior to determine what new content that user might like and then recommend that new content to the user.

Music isn't the only space where recommendation engines reign—Netflix and Hulu offer movie and TV recommendations, YouTube has its infamous right sidebar, and Amazon's website seems to fill every inch of extra space with recommended products. Even Facebook and Twitter recommend related accounts and pages to you or suggest "people you may know" to connect with.

At this point, recommendation engines are foundational to the appeal of a wide variety of digital services, that is, if they're useful. They can also be endlessly frustrating if they recommend irrelevant content or content they've already viewed. And these engines have a pretty short span with which to capture users' attention. Netflix research says that users will give up their search for new content to watch after 60 to 90 seconds.[31] If users are disappointed by the offerings of a **recommendation engine**, it may lead to a case of "failure runs deep," in which users may quickly learn that a recommendation engine is ineffective and begin to ignore it. But web services like Netflix and Spotify continuously tweak the algorithms to improve the suggestions.

At their best, recommendation engines serve an effective purpose for both parties. Web services want to keep users browsing, listening, watching, and shopping, and according to Netflix's research, the recommendation engine saves the company more than $1 billion by doing just that. Internal research claims that 80% of video views on Netflix come from recommendations

---

[31]McAlone, Nathan. "Why Netflix thinks its personalized recommendation engine is worth $1 billion per year." Business Insider. Last updated June 14, 2016. Accessed June 16, 2020. www.businessinsider.com/netflix-recommendation-engine-worth-1-billion-per-year-2016-6.

rather than from direct search.[32] Users want to find new and useful content, and recommendation engines help them do this.

In 2014, Netflix announced that they were updating their recommendation engine from a simpler engine focused on viewing data to a more complex one based on a neural network.[33] Netflix's recommendation algorithm uses data about users' viewing habits—including which shows and movies they have watched, how quickly they finished those shows, and which shows they quit watching. They combine this with data from other viewers, and with genre and feature codes assigned to each Netflix show by human coders, to group users into one of thousands of "taste groups".[34]

Spotify's recommendation engine for its Discover Weekly playlist features a similar apparatus of data collection and subgenre classification, but it adds in another metric: pairings on user playlists.[35] A song that many users have on the same playlist as the Cranberries' "Dreams" is pretty likely to be related to "Dreams" somehow. This further limits the chance of coincidental pairings. Spotify has also described personalizing its recommendations based on user habits. As of 2015, Spotify has described building a detailed database of minuscule subgenres much like Netflix's taste groups.

Spotify uses a three-pronged process to shape its Discover Weekly Recommendations, according to software engineer Sophia Ciocca. It runs a matrix analysis which compares users to other users, it uses natural language processing data from press coverage of artists and songs to determine what adjectives might describe them, and it runs a neural network analysis of the audio attributes of each of its songs.[36] It then combines these factors with human guardrails to avoid things like playlists full of children's music for parents.[37]

All of this personalization can give Discover Weekly the feeling of a human-curated playlist. In a sense, it *is* a human-curated playlist—it's created based on your listening habits and the playlists shaped by other users, and there is human

---

[32]McAlone, "Why Netflix."

[33]Russell, Kyle. "Netflix Is 'Training' Its Recommendation System By Using Amazon's Cloud To Mimic The Human Brain." Business Insider India. February 12, 2014. Accessed June 15, 2019. www.businessinsider.in/Netflix-Is-Training-Its-Recommendation-System-By-Using-Amazons-Cloud-To-Mimic-The-Human-Brain/articleshow/30259713.cms.

[34]Plummer, Libby. "This is how Netflix's top-secret recommendation system works." Wired. August 22, 2017. www.wired.co.uk/article/how-do-netflixs-algorithms-work-machine-learning-helps-to-predict-what-viewers-will-like.

[35]Pasick, Adam. "The Magic that Makes Spotify's Discover Weekly Playlists So Damn Good." Quartz. Accessed June 15, 2019. https://qz.com/571007/the-magic-that-makes-spotifys-discover-weekly-playlists-so-damn-good/.

[36]Ciocca, Sophia. "How Does Spotify Know You So Well?" Last updated October 10, 2017. Accessed June 15, 2019. https://medium.com/s/story/spotifys-discover-weekly-how-machine-learning-finds-your-new-music-19a41ab76efe.

[37]Pasick, "The Magic."

input throughout the process. But deep learning (which we might consider AI) plays a role—and, ultimately, your tracks are chosen by an algorithm.

One Spotify user, blogger Eric Boam, logged nearly all music recommendations he received for an entire year and wrote a post comparing Spotify's recommendations with those from media and human sources. He found that Spotify offered a large volume of recommendations, but that those recommendations had a lesser rate of success than recommendations from other humans or from the media.[38] So Spotify's algorithm doesn't quite have the quality of a human recommender just yet. Of course, it can spit out a far higher volume of recommendations.

And still, Boam describes engaging with Spotify recommendations and sometimes finding favorite albums through the service. Ultimately, engaging users with an AI service is the end goal, and recommendation engines achieve this. Discover Weekly will never replace your friends' recommendations, but it doesn't have to—those recommendations are probably shared in the form of a Spotify playlist anyway.

Recommendation engines exemplify the ways in which AI might fit into a user experience. While only a portion of Spotify's recommendation engine is in fact an AI system, that AI system blends seamlessly with other computing and human elements to build an engine that proves valuable to users. At the end of the day, this engine is only one part of the appeal of Spotify or Netflix as a service, but Netflix's numbers indicate that it is an important component of customer retention. Even if customers don't always get their music or movie recommendations from Discover Weekly or Netflix's recommendation service, users still engage with those digitally produced recommendations. And the high recommendation volume a digital recommender can produce can keep users engaging with a service before their attention spans expire.

The point   AI may not be the end-all, be-all that's ready to replace humans in fields like recommendation, but it can still be a useful component of a user's experience.

AI journalists

The field of journalism is in a unique predicament in the information age: despite being the most important purveyors of news in local and national settings, the news business is in a state of crisis. Newspapers, especially, have

---

[38]Boam, Eric. "I Decoded the Spotify Recommendation Algorithm. Here's What I Found." Medium. Last updated January 14, 2019. Accessed June 15, 2019. https://medium.com/@ericboam/i-decoded-the-spotify-recommendation-algorithm-heres-what-i-found-4b0f3654035b.

been ravaged by the digital era. Many smaller newspapers have shut down altogether, and those that have survived have often done so by laying off staffers and cutting pay, leading to a possible reduction of news quality. Journalists are already spread thin by the nature of the constantly churning news cycle, and they now have to complete a larger proportion of their paper's work on stagnant or reduced pay.

Media owners are searching for ways to cut costs, and journalists are looking to cut down on rote work. Enter AI journalists. In 2010, Northwestern University researchers released StatsMonkey, a program that could write automated stories about baseball games.[39] By 2019, major news outlets including *The Washington Post* and the Associated Press were using AI to write articles.[40]

Naturally, journalists are afraid of the potential of losing jobs to robots. But AI journalists are not quite capable enough to replace most journalists anytime soon. Generally, AI journalists do best at producing high volumes of short recaps on formulaic, data-based events like earnings reports and baseball games. AI journalists do not write articles all on their own, either—they're generally fed scripts on how to write a particular type of article.[41] The major AI journalism project RADAR operates much like an expert system, as journalists must program the format of articles on a certain topic into the system alongside a set of if-then rules.[42] Other AI journalists seem to operate similarly. While AI journalists can produce large amounts of content in the subfields for which they have been trained, they are generally not capable of replacing much of the cognitive work journalists do.

In 2019, the Tow Center for Digital Journalism at Columbia University released a study in which researcher Andreas Graefe attempted to use an AI journalist to automatically generate stories about polling and predictions in the 2016 US presidential election.[43] Graefe considered the project "very successful,"[44] as the AI published thousands of articles on different results. However, it was most successful with relation to specific domains for which it was meticulously trained and when dealing with less complex data.

---

[39]"Program Creates Computer-Generated Sports Stories." NPR. Last updated January 10, 2010. Accessed June 16, 2019. www.npr.org/templates/story/story.php?storyId=122424166.

[40]Peiser, Jaclyn. "The Rise of the Robot Reporter." Last updated February 5, 2019. Accessed June 16, 2019. www.nytimes.com/2019/02/05/business/media/artificial-intel-ligence-journalism-robots.html.

[41]Peiser, "The Rise."

[42]"Will AI Save Journalism—Or Kill It?" Knowledge @ Wharton, University of Pennsylvania. Last updated April 9, 2019. Accessed June 17, 2019. https://knowledge.wharton.upenn.edu/article/ai-in-journalism/.

[43]Graefe, Andreas. "Computational Campaign Coverage." Tow Center for Digital Journalism (2017). https://academiccommons.columbia.edu/doi/10.7916/D8Z89PF0/download.

[44]Graefe, 37.

Graefe found difficulty in training the AI to recognize such qualitative characteristics as one candidate having "momentum" or a margin being "large" or "small." In an extremely close race, a poll showing a three-point lead for one candidate might seem "large"; in a race with one dominant candidate, that same lead would be "small." Since these qualities vary in specific cases, they are difficult to quantify, making them difficult to program into the AI.

This is an effective illustration of the limitations of AI journalists today. They're essentially expert systems engineered for specific scripts and domains, while being unable to do the sort of adaptive or deep-learning work that might really make them a threat to human journalists. The sort of qualitative analysis that can easily be performed by humans is much more difficult for AI. Meanwhile, AI can mass-produce formulaic stories that human beings would otherwise have to individually write out. It leads to a symbiotic relationship.

In 2016, *The Washington Post* deployed AI in its election coverage. Even with the limitations of AI journalists, the *Post* successfully applied theirs to write some 500 election stories and internally to help alert reporters to unexpected shifts in election data and even won an award for their bot usage.[45, 46] In 2017, the *Post* hired a new team of human investigative journalists.[47] We think the timing of those two events may not be a coincidence.

AI AS A BEAT WRITER

GAVIN: If the *Post* can trust AI to fill out routine stories, they can afford to invest more in the sort of cognitive, investigative work that human beings are best at.

BOB: Which is exactly what we need to save journalism, right?

GAVIN: Right. *The Washington Post* can now afford to put additional resources towards in-depth reporting.

BOB: So, AI can do some of the routine work of journalism, writing repetitive stories about election polls and baseball games. That frees up resources for investigative reports.

---

[45]Moses, Lucia. "The Washington Post's robot reporter has published 850 articles in the past year." Digiday. Last updated September 14, 2017. Accessed June 17, 2019. https://digiday.com/media/washington-posts-robot-reporter-published-500-articles-last-year/.

[46]Martin, Nicole. "Did a Robot Write This? How AI Is Impacting Journalism." Forbes. Last updated February 8, 2019. Accessed June 17, 2019. www.forbes.com/sites/nicolemartin1/2019/02/08/did-a-robot-write-this-how-ai-is-impacting-journalism/#31e620777957.

[47]WashPostPR. "The Washington Post to create rapid-response investigations team." The Washington Post. Last updated January 9, 2017. Accessed June 17, 2019. www.washingtonpost.com/pr/wp/2017/01/09/the-washington-post-to-create-rapid-response-investigations-team/?utm_term=.5f4864546f4b.

GAVIN: What happens if the AI screws up? What if it says that my favorite baseball team, the San Francisco Giants, won by four runs, when they actually lost by four runs?

BOB: Well, people have a hard time blaming a computer. But I guess that speaks to the beauty of AI being limited to less consequential stories for the time being. If AI reports the wrong score for the Giants' game for a few hours, the world isn't going to burn down.

GAVIN: As a Giants fan, I'd be a little upset if I read a whole article about how they won, only to find out they actually lost. Maybe there should be some way of notifying me, as someone who read the article, about the mistake. Maybe they could send out a push notification to only users who opened the article, saying that there was an error.

BOB: I'm not sure the technology's there, but that's a good example of context-aware proactivity. For the time being, AI journalists are another example of AI supplementing human employees. AI can only write stories about specific domains, and they're writing the sort of stories that mainstream journalists might not enjoy. When it comes to more complex tasks, real journalists are better equipped to handle them. Fears of human journalists losing their jobs to AI are overexaggerated, as of now.

The point   AI in workplaces like the newsroom can actually elevate the work of human employees, allowing them to take on more cognitive tasks uniquely suited to humans' strengths.

AI, filmmaking, and creativity

In this chapter, we've tried to emphasize the fact that human beings and AI are good at different things. AI can make calculations, gather disparate sources of information, and find insights that human beings can't. But there are still plenty of domains where human beings still reign supreme. In order to find out how close AI is to beating us at our own game and venturing a look at how AI and humans can work together in more qualitative domains, we decided to take a look at how AI fares in the creative domain of filmmaking.

Believe it or not, the 2010s saw the first-ever AI-created movie trailer. In 2016, IBM's Watson partially constructed a trailer for *Morgan*, a horror film about humans dealing with an AI system gone rogue.[48] Watson was trained for the task by watching and analyzing other horror movies, to determine the different types of emotion present in each scene, and it then selected ten moments from *Morgan* that would fit well in a trailer, which were then

---

[48]Alexander, Julia. "Watch the first ever movie trailer made by artificial intelligence." Polygon. Last updated September 1, 2016. Accessed July 6, 2019. www.polygon. com/2016/9/1/12753298/morgan-trailer-artificial-intelligence.

compiled into a trailer by IBM.[49] The trailer has moments that seem scary or emotionally resonant, and it has the right mood-setting music, but the trailer is also disjointed. The different clips don't always fit well together, and it's hard to tell what the plot of the film is based on the trailer. The AI-generated *Morgan* trailer has an "uncanny valley" quality to it—that is, it feels close enough to a human-created trailer that it isn't totally incomprehensible, but just different enough that it leaves the viewer with an eerie effect.

Film editing combines cognitive and affective elements, requiring the editor to understand what the emotional resonance of a scene or clip will be and how best to design a trailer or a film to maximize the sort of emotion that they want to evoke. It's these sorts of tasks, which are both complex and emotional, that AI is least equipped to take on. Consider Sherry Turkle's criticism of the Hello Barbie toy: AI can't provide empathy.[50] Empathy is a key component of a lot of art.

This clearly illustrates what human beings have that AI is a long way away from being able to replicate. Humans are capable of constructing coherent narratives that make sense to other human beings, evoke emotions in their audiences, convey subtextual messages, and even contain aesthetic beauty. AI can't do any of those things. It's difficult to quantify aesthetics.

So, AI is pretty bad at scriptwriting and not great at generating movie trailers, either. But it's still proven useful to filmmakers in the realm of animation. AI has made the routine tasks of animation—fixing little details in a character's movement or definition—much easier. In today's special-effects-heavy films, this is a vitally important contribution. Today's actors often star as characters whose physically impossible appearance needs to be animated. At one time, this required these actors to film all of their scenes in front of a green screen or inside a recording studio. Today's AI can artificially generate a character's look based on the actor's face and allow them to act with the other actors in the film, animating the character on top of their real-life appearance. This technology was used in a recent *Avengers* movie. It can even generate the animated character so quickly that the actor can view themselves acting as their animated character while filming.[51]

[49]20th Century Fox. "Morgan | IBM Creates First Movie Trailer By AI [HD] | 20th Century FOX." YouTube. Published August 31, 2016. Accessed July 6, 2019. www.youtube.com/watch?v=gJEzuYynaiw.

[50]Barbie wants to get to know your child. Vlahos, James. Sept 16, 2015. Retrieved May 19, 2020. nytimes.com/2015/09/20/magazine/barbie-wants-to-get-to-know-your-child.html

[51]Robitzski, Dan. "Was That Script Written By A Human Or An AI? Here's How To Spot The Difference." Futurism. Published June 18, 2018. Futurism. Accessed July 6, 2019. https://futurism.com/artificial-intelligence-automating-hollywood-art.

It's not just filmmaking where AI has made an impact in automating rote tasks. AI-human team relationships not unlike the one found in medical AI are all over the artistic world. In visual art, the Celsys AI can colorize black-and-white drawings.[52] In music, the Bronze AI can generate an infinite version of an individual song that changes a little bit every time it's played.[53] In fiction, one author has created an AI program that automatically auto-completes a writer's sentences while writing science fiction stories, based on a corpus of science fiction stories.[54] He envisions the program as a kind of co-author, which generates ideas that might spark the writer's human creativity. Tellingly, none of these three projects are widely used. AI in the arts is not quite ready for prime time yet.

But these projects offer a preview of how AI might begin to make inroads into those domains that are very conducive to human skills and not so conducive to the quantitative expertise of AI. In almost any domain, there is room for a useful assistant that brings a different perspective and a contrasting set of skills. In the arts, there may be fewer quantitative tasks for AI to complete, but there are still specific domains where it can automate tasks that once took humans hours of frustrating effort to complete, freeing them up to complete other tasks.

■ **The point**　In the arts, human skills still reign supreme. But artists are already finding niches where AI can offer a contribution, bringing the collaborative relation to even the more qualitative domains.

Business AI

Business is one of the AI verticals where human-AI relationships will be more difficult to forge. There are AI solutions available for finding new sales leads, analyzing job applicants, improving customer service, parsing turgid legal

---

[52]Lee, Dami. "AI can make art now, but artists aren't afraid." The Verge. Last updated February 1, 2019. Accessed July 6, 2019. www.theverge.com/2019/2/1/18192858/adobe-sensei-celsys-clip-studio-colorize-ai-artificial-intelligence-art.

[53]Christian, Jon. "This AI Generates New Remixes of Jai Paul...Forever." Futurism. Last updated June 4, 2019. Accessed July 6, 2019. https://futurism.com/the-byte/ai-remixes-jai-paul.

[54]"Writing with the Machine." robinsloan.com. www.robinsloan.com/notes/writing-with-the-machine/.

documents, and more.[55, 56, 57] But with their profits riding on it, it's difficult to imagine businesspeople abandoning their tried-and-true sales and hiring methods and adopting one of the various AI business solutions on the market. It's going to require building an AI system that is especially capable of building trust in a business setting.

Luckily, there is already research about what it takes to build an effective business relationship, thanks to guidelines designed for human-to-human collaboration. We think this research is a good starting point for building a relationship between AI and humans in the workplace. An important contribution on the subject comes from the medical field of family practice, but its findings are applicable across many types of businesses. Research from FPM, an American Academy of Family Physicians journal, identified seven elements that lead to healthy, collaborative relationships: trust, diversity, mindfulness, interrelatedness, respect, varied interaction, and effective communication.[58]

Six out of seven of these elements are vitally important to human-AI interactions in the business field, and we've already discussed three of them. Beginning in Chapter 1, we've extensively discussed the importance of trust for AI. We've also discussed the manner in which AI offers diversity by investigating the collaborative relationship between AI and human beings. AI can add diversity of perspective to a decision-making process because its computational nature frees it from cognitive biases inherent in human reasoning. Interrelatedness is similar to context—it requires an awareness of the big-picture significance of specific actions and collaborators within an organization.

But it also might make users more comfortable collaborating with a machine, by making the machine seem more friendly and less intimidating.

This is another area where Microsoft's Cortana offers a unique approach that makes its stagnation in the wake of the virtual assistant winter all the more frustrating. At one point, after Microsoft purchased LinkedIn in 2016, they

---

[55]Power, Brad. "How AI Is Streamlining Marketing and Sales." Harvard Business Review. Last updated June 12, 2017. https://hbr.org/2017/06/how-ai-is-streamlining-marketing-and-sales.

[56]"Applications of Artificial Intelligence Within your Organization." Salesforce. www.salesforce.com/products/einstein/roles/.

[57]Greenwald, Ted. "How AI Is Transforming the Workplace." The Wall Street Journal. Last updated March 10, 2017. www.wsj.com/articles/how-ai-is-transforming-the-workplace-1489371060.

[58]Tallia, Alfred F., Lanham, Holly J., McDaniel, Jr., Reuben R., Crabtree, Benjamin F. American Association of Family Practitioners. "Seven Characteristics of Successful Business Relationships." From *Fam Pract Manag.* 2006: 13(1):47–50. Accessed July 14, 2019. www.aafp.org/fpm/2006/0100/p47.html.

were rumored to be planning to bring its data to Cortana.[59] This could have created an AI assistant that truly had the capability to function in the realms of both business and personal affairs. Unfortunately, this version of Cortana never came to fruition.

The FPM researchers describe effective communication as knowing when to apply two different types of communication: text communication (which delivers less information and context but is quick and convenient) and face-to-face or over-the-phone communication (which is more informative and context rich, but cumbersome). Business AI might apply similar methods in deciding how to communicate a message to the user. While the option of face-to-face communication is not present for AI, there are several methods of possible communication between AI and humans, including typing back and forth, selection from a menu, and voice interfaces. An effective business AI might be built with information about which ones to use when.

R-E-S-P-E-C-T WHAT IT MEANS TO ME

BOB: That leaves respect as the final one of our seven elements of business communication.

GAVIN: How does respect differ from **trust**?

BOB: Well, usually, when we respect someone, we also trust them. So, it can be hard to separate them. But in my mind, respect is something above and beyond trust. You might trust a business AI tool to do one or two specific things, but not respect it. I might trust my AI to analyze my earnings report and write up some statistical insights, but that doesn't mean I respect it. The FPM researchers talked about respect as "valu(ing) each other's opinions." To me, that means that you think of its insights as worth considering in any case. If an AI system that I respect tells me about some insight—even if that insight goes against all my preconceived notions—I'll listen to it. I'll hear it out.

GAVIN: Respect is more general, and trust is more specific?

BOB: Right. That's how I would differentiate it. Although it's difficult, because we sometimes use "trust" to mean what I would call respect. But right now, AI is built to value your opinion, but I might not value its opinion, even if I trust it to do certain things. AI has to do all of these other things—earn trust, apply context, communicate effectively, and maybe even vary interactions—before it can earn the user's respect.

---

[59]Darrow, Barb. "How LinkedIn Could Finally Make Microsoft Dynamics a Big Deal" June 13, 2016. https://fortune.com/2016/06/13/microsoft-linkedin-dynamics-software/ (accessed May 22, 2020).

GAVIN: A lot of that sounds like UX. AI needs to be designed for its users, with respect to how they think and the components that make them trust. What makes you trust another human being isn't all that different from what makes you trust AI.

BOB: And that respect is especially hard to earn in a business setting. There are serious consequences if the AI gets anything wrong.

GAVIN: An element that's not on this list—but maybe should be, because it's a component of respect—is transparency. I need to have some idea of what AI is doing, and why it's doing it, before it can earn my respect. I don't need to know all the details, but I need to know something.

BOB: Sounds like something to add to our AI-UX framework.

The point A good user experience can help build trust and respect in the AI application.

Some conclusions and where we're going next

In the next chapter, we will go deeper into how to affect change in the areas of AI that we can influence, such as the data. Again, the focus is not on a particular AI learning algorithm or code, but what can we do to improve AI through the data itself? What are the elements where we can have an impact on AI? As we identify gaps and concerns, what can be done to pivot problems into opportunities to make the product better? What can we learn from those who understand the issues with datasets, for example, and how the "big players" in the industry approach solutions? In some ways, the answer is not simply to do what is needed, but to go orders of magnitude greater to give their AI-enabled products a better chance for success.

Garbage In, Garbage Out

Doing a disservice to AI

Given the ever-evolving nature of AI, programmers need to continuously improve and refine their algorithms. In Chapter 1, we saw how algorithms are improved and often repurposed for different tasks, such as the credit card fraud detection system called Falcon that Craig Nies described had its roots in a visual system to detect military targets. Essentially, the foundation for pattern recognition to differentiate battlefield equipment from surrounding landscapes was applied to recognize patterns of fraud in credit card data.

But, again, let's assume the AI code works; that is, the AI algorithms that feed all of the modern-day AI systems—whether they are called deep learning or machine learning or some other proprietary name—are capable of doing the job. If this is true, then the focus shifts from the code to the datasets that feed these systems. How much care has been placed into the data that feeds the machine?

We need to take a bit of a step back here and more clearly define the space we're talking about. AI is a huge field. The focus of our data-centered discussion here points at AI-enabled products that rely on human behaviors, attitudes, opinions, and so on. Data that are actively solicited (e.g., surveys) have

© Gavin Lew, Robert M. Schumacher 2020
G. Lew and R. M. Schumacher, *AI and UX*,
https://doi.org/10.1007/978-1-4842-5775-3_4

different properties (and problems) than data that are passively acquired. A lot of our discussion in this chapter focuses on data that is intentionally acquired from people.

WHAT WE FEED THE ALGORITHMS MATTERS

BOB: Consider Formula One racing. No matter how good the engine is, success depends on the entire ecosystem around the vehicle: the mix of the fuel, the skill of the driver, the efficiency of the pit crew, and so on.

GAVIN: An engine using low-grade fuel will underperform. In the case of AI, its fuel is data. While data scientists can massage the data to map to the algorithms for learning, how much care is placed on the dataset? That data might have been purchased from a site where it was "not perfect, but good enough." Or it may be far removed from the researchers who collected it. What if the data is no longer high grade?

BOB: As UX researchers, we know a lot about data collected from people—it's messy—the nuances in the questions, missing cells, context of collection, and more. The problem can be especially concerning if the dataset was not commissioned by the team using it. There's a lot of trust that the dataset is clean.

GAVIN: AI algorithms are fed data initially to learn and train those models; those models are then applied broadly to more data to provide the insight.

BOB: The data that is fed into AI is pretty important for success, especially in the training phase when AI is learning.

The point   It's not just how good the algorithms are; it's how good the data is or "garbage in, garbage out." Let's spend time giving AI the best data we can.

Swimming in data

As researchers, we (the authors) often talk to companies about their data. We ask about what they know, what they don't know, and what is currently being collected. We look for gaps and opportunities where more or better data could answer strategic questions. A common issue is that they have more data than can be analyzed. So it's not about collecting more data, but taking the time to think through how to better analyze what they have.

If this is the case, then the product team developing the AI-enabled technology must look critically into what data is used for training and what is used once the algorithm is trained. AI is an ecosystem of many elements—the algorithm is just one piece. It may be at the center and gets much of the attention, but success depends on all the elements being aligned and supportive of the objectives.

Companies swimming in data should think about how their data was obtained—did it come from a vast warehouse of compiled data, or was it gathered for a specific purpose? This is a critical question to help understand that data. Was the data specifically collected for the AI-enabled product that targets the key area of interest? If the data was not commissioned specifically for AI, then one must spend time to understand more about the dataset itself.

Questions to ask when evaluating a dataset are as follows:

- Where did the dataset come from?

- What was the method of data collection?

- If it was survey data, what are the assumptions and conditions under which this data was obtained?

- Were any of the data imputed (missing cells filled algorithmically)?

- What other datasets could be joined to add supplemental context?

- What do subject matter experts know about the data and how could this knowledge be beneficial to learning?

The point   These simple questions can identify areas of improvement for the training dataset that will be used to help AI learn. This is where we give AI a fighting chance at success by potentially giving the data more **context**.

So, how does AI really "learn"?

Data that capture human behaviors and interactions are given to machine learning (ML) scientists to train AI systems and algorithms. Whether the data comprises a set of liver-disease diagnoses and outcomes, comes from a consumer survey on attitudes toward marijuana usage, or derives from active/passive data collection of spoken phrases, AI systems need training data to ensure their algorithms produce the right outcomes. Custom-built data for AI may not be as common as datasets that were created for other purposes, such as market research, customer segmentation, sales and financial data, health outcomes, and a lot more. Once ML scientists have acquired a dataset, they still need to consider whether it includes what the AI system needs.

An example of how AI learns

At one level, AI can be thought of as a pattern recognition system. In order for an AI system to learn, it needs lots of examples. AI algorithm needs data to look for patterns, make mistakes, and refine its internal understanding to be better. As a fun example of this, Figure 4-1 illustrates an Internet meme that circulated a few years ago. What's interesting is that it shows how easy it is for people to detect the signal (the Chihuahua) in a very noisy field. AI algorithms have a very difficult time with this and these samples are useful to validate patten recognition systems.

Figure 4-1. Example of the challenge of pattern recognition and data that might be provided for an AI to learn how to distinguish between a Chihuahua from a blueberry muffin.

Different ways machines learn today

In general, there are three kinds of machine learning (ML) techniques for constructing AI systems, as follows:

- **Supervised learning** – in this approach, scientists feed algorithms a dataset comprising data—for example, labels, text, numbers, or images—and then calibrate the algorithm to recognize a certain set of inputs as a particular thing. For instance, imagine feeding an algorithm a set of pictures of dogs, in which each picture contains a set of features that correspond to properties of the picture. Inputs to the algorithm could also include a number of images that are *not* of dogs—for example, pictures of cats, pigeons, polar bears, pickup trucks, or snow shovels—and the corresponding properties of each of the *not-dogs* images. Then, based on what the algorithm has learned about classifying images as *dog* or *not dog* through the features and properties of images, if you show the algorithm a picture of a dog it's never seen before, it has the ability to identify that it is, in fact, a picture of a *dog*. The algorithm is successful when it can accurately recognize an image as a dog and reject images that are not dogs.

- **Unsupervised learning** – this approach attempts to find classes of similar objects in a dataset based on each object's properties. When scientists give an algorithm a set of inputs that have parameters and values, it tries to find common features and group them. For example, scientists might feed an algorithm thousands of pictures of flowers with various tags such as color, stem length, or preferred soil. The algorithm is successful if it can group all flowers of the same type.

- **Reinforcement learning** – this approach trains an algorithm through a series of positive and negative feedback loops. Behavioral psychologists used this technique of feedback loops to train pigeons in lab studies. This is also how many pet owners train their animals to follow simple commands such as *sit* or *stay* and then reward them with a treat or reprimand them with a *no*. In the context of machine learning, scientists show an algorithm a series of images, and then as the algorithm classifies images—of, say, penguins—they confirm the model when the algorithm properly identifies a penguin and adjust it when the algorithm gets it wrong. When you hear about bots on Twitter that have gone awry, this is typically an example of reinforcement learning where the bots have learned to identify examples incorrectly, but the system thinks they are correct.[1]

Although all ML techniques are useful and applicable in various contexts, let's focus on supervised learning.

---

[1] https://www.theverge.com/2016/3/24/11297050/tay-microsoft-chatbot-racist.

All data are not equal

Obtaining good training data is the Achilles heel of many ML scientists. Where does one get this type of data? Getting data from secondary sources is surprisingly easy. There are many sources[2] that provide access to thousands of free datasets. Recently, Google launched a search tool to make publicly available databases for ML applications easier to find. But it is important to note that many of these databases are very esoteric—for example, "Leading Anti-aging Facial Brands in the U.S. Sales 2018."[3] Nonetheless, data is becoming more accessible. While this supports educational endeavors, the ability for businesses to use these databases for mainstream applicability may be low.

These databases have limitations such as the following:

- They might not have precisely what ML researchers are seeking—for example, videos of elderly people crossing a street compared to children riding bicycles.

- They might not be tagged appropriately or usefully with the metadata that is necessary for ML use.

- Other ML researchers might have used them over and over again.

- They might not represent a rich, robust sample—for example, a database might not be representative of the population.

- They might lack enough cases/examples.

- They might not be very clean—for example, they could have lots of missing values.

As many researchers often say, all data are *not* equal. The inherent assumptions and context that are associated with datasets often get overlooked. If scientists do not give sufficient care to a dataset's hygiene before plugging it into an ML system, the AI might never learn—or worse, could learn incorrectly, as we described earlier. In cases where the quality of the data may be suspect, it's difficult to know whether the learning is real or accurate. This is a huge risk.

---

[2]https://medium.com/towards-artificial-intelligence/the-50-best-public-datasets-for-machine-learning-d80e9f030279.
[3]www.statista.com/statistics/312299/anti-aging-facial-brands-sales-in-the-us/.

Knowing what we now know about machine learning and the risks and limitations of datasets, how can we mitigate these risks? The answer involves UX.

PLAYING CATCH UP WITH COMPUTING SPEED WHEN WE SHOULD BE SLOWING DOWN

BOB: Recovering from failure and learning is necessary and part of how AI will evolve and succeed. But, recovering from failure requires significant overhaul to not do what did not "work" but review what "worked" as well.

GAVIN: Yes. And consider how fast the technology is advancing. The faster AI advances, in some ways, we lose the opportunity to think about ethical considerations or even about revisiting the foundations.

Consider the evolution of the CPU. Under Moore's Law, the number of transistors in a CPU doubles every 2 years, but in AI's case, computing power for AI took advantage of the massively parallel processing of a GPU (graphics processing unit). These are the new graphics chips associated with making video games smoother and the incredible action movies we see today. Massively parallel processing required to present video games made AI much, much faster. AI systems often took months to learn the dataset. When graphics chips were applied to AI applications, training intervals dropped to single days, not weeks. There is barely enough time to stop and think about the results.

BOB: If AI applications are to learn, they learn from consuming data. When one thinks of data, it is easy to assume that the AI application would take all data into consideration. But, practically speaking, consuming data still takes time. If the training sets are consumed faster, have we evolved our thinking on the data itself or just the processing power?

GAVIN: This is the trap. With all the emphasis on hardware and algorithmic advances, my fear is that this only distracts from getting the foundation right first.

BOB: Simply put, "garbage in, garbage out." We are doing a disservice to AI if we don't think about what we are feeding it.

The point   Advances in AI will come, but are we taking time to understand the data that we feed into the machine?

Getting custom data for machine learning

While not all datasets relate to human behavior, the majority of them do. Therefore, understanding the behaviors that the data capture is essential. Over the last decade, our UX agency has been engaged by many companies to collect data for custom datasets. This means we had to collect precise examples and attribute tags that are necessary to train or prove in their AI algorithms. (In some cases, thousands of data points are needed, which are samples of different things.) Here are some examples of these samples:

- Video samples of people doing indoor and outdoor activities

- Voice and text samples of doctors and nurses making clinical requests

- Video samples of people stealing packages from porches

- Video samples capturing the presence or absence of people in a room

- Thumbprint samples from specific ethnic groups

- Video and audio samples of people knocking on doors

Note that *none* of this data was available publicly. We had to build each of the datasets through custom research based on the specific intentions and research objectives of our clients.

CUSTOM DATA FOR AI IS A BIG DEAL

BOB: When we received a request from a client to collect thousands of samples for a custom dataset, we raised an eyebrow at how to approach this from a practical perspective. Thousands of people—face to face data collection! We were to capture *in situ* behaviors.

GAVIN: After reviewing the specifications, the amount of precision required was immense. Because participant demographics are always important to ensure samples are representative to the target population, we would need many participants. For instance, a facial recognition AI on a smartphone or computer needs to learn to recognize data from the same participant in different situations. They might have changed their appearance. So, we would collect data with and without beards. They could wear different clothing or different makeup, or different hair styles, and so on. We would systematically ask participants to change their look to add additional samples to the data. This would allow AI to learn about people, but also train it to recognize that the same person can look different. We were asked to capture participants in a variety of contexts. The amount of care placed on the ask was well thought through.

BOB: It also extended to different continents too. At some point, we argued that there were more cost-effective ways to collect this massive amount of data than through a UX firm like ours, as we tended to collect data on a much smaller basis. The response was that they understood, but that most large-scale data collection lacked the experimental rigor to capture what was needed for their AI application. They wanted to use precision that was typically used in small sample research studies but replicated two orders of magnitude higher.

GAVIN: When we write about using datasets that were used for other projects or datasets that are purchased and used for AI, the difference between that and commissioning a custom dataset for a specific AI application is striking. It is one thing to "dust off" an old dataset and entirely another world to specify what your AI would need to consume to properly train.

The point   While an existing dataset might describe people and behavior, a custom dataset can be tuned to the elements that make AI smarter and better. Care for the details in the data makes AI better.

Understanding the sheer magnitude of data collection necessary for effective ML applications, it seems obvious that these datasets should be custom. But, how much time, effort, and money are *actually* spent on clean datasets relative to the programming?

For many scientists and researchers, the easy way is to use data that already exists. But our clients who commissioned these projects understood a key shortcoming of these methods: low data integrity. The project sponsors recognized that the underlying data had to be clean and representative of the domain they were trying to model—carefully considering the nuances of captured experiences. So, we needed to collect the behaviors in context and had to *observe* them—not simply ask for a number on a five-point scale—as is often the case in quantitative data collection. Apart from the obvious problems in survey research, we, as psychologists, understand that people often cannot report on their own behaviors reliably. That is, we can't often just ask people to tell us what they did; we must observe and record. Capturing behavior is the prerogative of UX and requires research rigor and formal protocols. What we learned is that UX is uniquely positioned to collect and code these data elements through our tested research methodologies and expertise in understanding and codifying human behavior.

Data hygiene

While this section may seem somewhat redundant with some of the material covered in Chapter 3, the dataset can be fraught with concerns, such as the following:

- Identification of missing cells that are filled with imputed data where cell tags (i.e., notations that this is imputed) are not passed on to the AI team

- Data that is purposefully and systematically unsurveyed so multiple participants are combined to complete a full survey (i.e., split questionnaire survey design).

- Surveys that are completed by bots, acting as humans[4]

Data scientists take data hygiene very seriously. What we are concerned with here is that we have heard from those involved immersed in AI development that sometimes the preceding issues are glossed over. Let's not assume the data is free from elements that can skew the learning.

The point   Let's not assume the data—even if no cells are missing—is free from elements that can bias outcomes and create unintended consequences.

DOING A DISSERVICE TO AI

BOB: The question for anyone who is working on AI applications is how much care is spent on the data itself?

GAVIN: This is a challenge because so many hands touch the data and when it is passed from survey designers to programmers to respondents to data scientists then to AI technologists, who know exactly what was done to the dataset that might be an artifact that will influence the AI application?

BOB: We know that some data scientists tag fields that have been imputed, but by the time the data is washed and formatted for training the AI application, has this knowledge been stripped away?

GAVIN: There is a disservice to AI to have it trained on datasets where there may be underlying flaws in the data..

---

[4]In fact, in one study we were involved in, as many as 10% of the respondents were estimated to have been from bots built to take the survey to collect the incentive.

The point   The dataset deserves a lot of scrutiny—ask questions on the methodology, respondents, questions, design, and so on. This is what the AI application will use to learn and all team members can play a role to give AI better data.

Black box implications

As described in Chapter 3, one challenge with AI is that it does not reveal the meaning or rationale behind what it finds. It is a classic "black box" where data goes in and an answer comes out, but there is no description of why or how the answer came to be.

As mentioned above, potentially compounding the problem is the concern that the data we think is obtained from humans, just might be from bots. Or in our efforts to make a complete dataset, we use imputed data where an equation or algorithms were used to fill data elements. The concern is that any outcome obtained might simply be the result of the AI system reverse engineering the imputation algorithms used. Because AI is a black box, we are not able to inspect the "why" behind AI results. This takes away our ability to walk backward through the AI application's conclusions to find the underlying rationale. This can be problematic, especially considering how fast the business world acts on AI findings.

ETHICS IS BEST EARLY NOT LATE IN THE DISCUSSION

BOB: When I started my career, there were companies that were considered innovators and there were those that adopted a "wait and see" attitude about "fast following" of innovations.

GAVIN: Today, these corporate philosophies still exist, but it seems that the brand value of being innovative is much stronger, and this is driving companies to innovate faster and faster. Consider the practice of producing the minimum viable product (MVP), where startups and monolithic companies alike are launching products with a bare minimum feature set hoping to capture the attention of the marketplace and learn quickly from customers.

BOB: One challenge of MVP is what happens if the product has been pared down such that isn't very compelling in its MVP state? This is not only a UX and value proposition issue, but what concerns me more about AI is how quickly companies are moving to be first to market. Let's say you are creating an AI-enabled application. You rush to get data from sources that make sense. The dataset is cleaned and used for training. After AI "trains" and presumably "learns," it identifies an interesting finding. What does a company do next?

GAVIN: A company that believes they are "innovative" will run to build a business case, get funding, and build a product where AI is at the core. But what if the dataset is dodgy or biased due to poor sampling?

BOB: You are talking about ethics in data. This is an area where AI has not developed fully. Companies are building AI not for foundational science, but for commercial advantage. The same sorts of issues that arise with bias in the culture also exist in the data. So the fear is that AI applications may have subtle—or even not-so-subtle—biases because the underlying data contain biases.

The point Organizations are moving fast to build applications, but social and ethical considerations inherent in the data need to be addressed, developed, and adopted.

Next, let's take a deeper look at ethics and AI through the lens of privacy and bias.

Ethics and AI

The ethics of AI is a relatively new area of discussion. Only recently has AI become mainstream enough that ethical considerations are beginning to take shape. There are no formal ethical standards or guidelines for AI. It is very much the proverbial "Wild West" where technology is being created without guardrails.[5]

The concern is that the "grist for the AI mill" (the data) could hide ethical concerns. What data was used? Was the data universal? Did it have too much focus on a region or socioeconomic level? If the training data contain bias, would AI have an opportunity to revisit the underlying training or will it always have a bias?

Let's look at two important points concerning ethics and AI: privacy and bias.

Privacy

The data science revolution is the centerpiece of major tech companies. As outlined by Facebook investor Roger McNamee, web startups like PayPal, Facebook, and Google have made massive inroads through a big-data first approach—using data to build more functional and successful products, then

---

[5]There is some traction emerging: in June 2019, Oxford University announced a donation of £150M to establish the Schwartzman Centre that will contain the Institute of Ethics and AI. www.ox.ac.uk/news/2019-06-19-university-announces-unprecedented-investment-humanities (retrieved March 1, 2020).

selling that data.[6] Despite being well connected in the tech industry, McNamee sounded the alarm about tech companies' big-data focus. He invoked the idea that user privacy is actively being compromised by major tech firms in a way that outweighs any benefits of their services. While privacy concerns haven't stopped programs like Gmail and Facebook from becoming behemoths, they are an ever-present part of the discussion around issues of big tech, and AI may only exacerbate these fears. In 2010, then-Google CEO Eric Schmidt described Google's capabilities in terms sure to scare any user concerned about their privacy:

> *We don't need you to type at all. We know where you are. We know where you've been. We can more or less know what you're thinking about.*[7]

This quote from a decade ago described how algorithms guided by fallible human beings could extract untold insights from the data we all share online. When Eric Schmidt was asked during an interview for CNBC's "Inside the Mind of Google" special about whether users should be sharing information with Google as if it were a "trusted friend," Schmidt responded:

> *If you have something that you don't want anyone to know, maybe you shouldn't be doing it in the first place.*[8]

When you consider what is in a dataset and how it is derived from human behavior, this is a clear example of behavior and how it can be used to analyze and predict future behaviors. The message that Schmidt might not be explicitly describing is how much information Google really has. It is certainly more than simply search terms, but geonavigation data, actual consumer purchases, and email correspondences at the very least. And the rub of it all is that we give our consent to have this data collected by clicking through and accepting the policies. We're all giving up our privacy for the putative benefits that the technology offers us.

---

[6]McNamee, Roger. "A Brief History of How Your Privacy Was Stolen." The New York Times. June 3, 2019. Accessed June 3, 2019. www.nytimes.com/2019/06/03/opinion/google-facebook-data-privacy.html.

[7]Saint, Nick. Google CEO: "We Know Where You Are. We Know Where You've Been. We Can More Or Less Know What You're Thinking About." Business Insider. October 4, 2010. Accessed June 25, 2019. www.businessinsider.com/eric-schmidt-we-know-where-you-are-we-know-where-youve-been-we-can-more-or-less-know-what-youre-thinking-about-2010-10?IR=T.

[8]Esguerra, Richard. "Google CEO Eric Schmidt Dismisses the Importance of Privacy." Electronic Frontier Foundation. December 10, 2009. Accessed February 16, 2020. www.eff.org/deeplinks/2009/12/google-ceo-eric-schmidt-dismisses-privacy.

Privacy can be divided into three different types:

- Big Brother privacy (keeping personal information from government or business entities)

- Public privacy (keeping personal information from coworkers or community)

- Household privacy (keeping personal information from family or roommates)

Each of these three types of privacy has different impacts on UX.

For a long time, Big Brother privacy intrusions have mostly been tolerated by users. After all, we've all had the experience of clicking through the Terms and Conditions for a new account or app without reading them. But, with the era of big data fully upon us, the issue seems to be gaining political salience. This is best exemplified by the European Union's GDPR privacy law, one of the most prominent attempts to regulate big data. The GDPR is "based on the concept of privacy as a fundamental human right."[9] Privacy and policy research director Michelle Goddard views the GDPR's regulations on data collection as an opportunity for data scientists, not a setback. She says the GDPR's focus on ensuring privacy through "transparency" and "accountability" aligns with privacy practices necessary for ethical research, including anonymizing personal data.[10] AI, similarly, can focus on transparency to dispel user concerns about Big Brother privacy.

Public privacy is probably the least likely of these three forms to be violated given the current political and mainstream concerns focused on big businesses like Google and Facebook, so let's look at household privacy.

Household privacy is most salient with programs or devices that are meant to stay at home or to be used by one user in particular, such as standalone virtual assistants. If a user buys a virtual assistant device for their household, it can lead to violations of household privacy. For example, the user's roommate might be able to read and respond to their texts, or their spouse might stumble upon an update on the delivery status of their secret anniversary gift. The desktop computers of a bygone era were a classic case of potential household privacy violations, which were resolved by the feature of individual user profiles. A similar solution might help virtual assistants—but the technology for a convenient profile solution on virtual assistants is still evolving.

---

[9]Goddard, Michelle. "The EU General Data Protection Regulation (GDPR) is a European regulation that has a global impact." International Journal of Market Research 59/6 (2018). https://journals.sagepub.com/doi/10.2501/IJMR-2017-050.
[10]Goddard, "The EU."

Mattel created a virtual assistant that offers a glimpse at a profile system. Aristotle was Mattel's virtual assistant, based on Amazon Alexa, which was intended to primarily serve children. The company planned to make Aristotle capable of understanding a child's voice and of differentiating it from adult voices. Then, the device could offer limited capabilities to child users, while also offering adults the ability to use Alexa to do more complex tasks like ordering childcare supplies.[11] However, Aristotle was canceled in 2017, after consumer advocates, politicians, and pediatricians objected. Big Brother privacy concerns were one major reason for objections to Aristotle, along with concerns about child development.[12]

While Aristotle may not have come to fruition, an AI system like it that can differentiate users' voices from one another and associate with an individual's profile is a solution to the problem of household privacy in virtual assistants. There are other possible solutions, of course—perhaps a future assistant could determine who it is talking to by discovering whose smartphone is in the room. In 2017, Google Home provided a feature where it could distinguish from up to six different household members[13] and Amazon's Alexa followed suit in 2019 with "Voice Profiles."[14]

Users' expectations of privacy online can be slippery, as Microsoft principal researcher Danah Boyd has pointed out. Boyd has written that users' expectations of privacy online are most obviously violated when the context is stripped away from their actions and those actions are released to a wider public than the user had intended. This leads the user to feeling a loss of "control over how information flows,"[15] which results in user mistrust in the technology that removed the context.

For an example of how to build trust, let's turn back to Spotify. The company cites data claiming that it is more trusted than its competitors, including among millennials. They cite "discovery" features like Discover Weekly and

[11]Wilson, Mark. "Mattel is building an Alexa for kids." Fast Company. January 3, 2017. Accessed June 25, 2019. www.fastcompany.com/3066881/mattel-is-building-an-alexa-for-kids.

[12]Vincent, James. "Mattel cancels AI babysitter after privacy complaints." The Verge. October 5, 2017. Accessed June 25, 2019. www.theverge.com/2017/10/5/16430822/mattel-aristotle-ai-child-monitor-canceled.

[13]Baig, Edward C. "Google Home can now tell who is talking." USA Today. April 20, 2017. Accessed February 16, 2020. www.usatoday.com/story/tech/talkingtech/2017/04/20/google-home-can-now-tell-whos-talking/100693580/.

[14]Johnson, Jeremy. "How to setup Amazon Alexa Voice Profiles so it knows you are talking." Android Central. November 26, 2019. Accessed February 16, 2020. android-central.com/how-set-amazon-alexa-voice-profiles-so-it-knows-its-you-talking.

[15]Boyd, Danah. "Privacy, Publicity, and Visibility." 2010. Microsoft Tech Fest, Redmond, WA. Accessed June 4, 2019. www.danah.org/papers/talks/2010/TechFest2010.html.

the partially neural-network-powered recommendation engine as a primary reason why.[16] In an article directed at advertisers, Spotify claims that users are willing to give a company personal information so long as it results in a useful feature. Spotify's recommendations are that useful feature.

The Spotify recommendation engine is built based only on data from Spotify itself, and it even allows users to enter a private mode in which Spotify won't count their streams. That means that users can simply take their guilty pleasures elsewhere (might want to stream that Nickelback album in private mode or on YouTube) and make sure they don't affect their recommendations. This helps users trust that Spotify's data collection serves a purpose for them.

AI DOES NOT KNOW WHERE THE "LINE" IS, SO WE NEED TO DRAW IT

GAVIN: This is a very difficult question to solve for businesses because companies have an obligation to their shareholders first, so AI-enabled products should be made with all data available to produce a compelling product.

BOB: But, if users rebel, that will hurt the shareholders. Companies must still balance privacy to not negatively impact their brand.

GAVIN: This reminds me of another quote by Eric Schmidt. When he was asked about whether Google would implant technology into the brain to get information, Schmidt said, "There is what I call the creepy line. The Google policy on a lot of things is to get right up to that creepy line and not cross it."[17]

BOB: And let's hope we can trust businesses to know where that line is.

The point　The need for AI to respect privacy comes from those who develop and market AI.

Privacy implications center around whether the data should be used in AI. Let's explore the concept of bias that creeps into our dataset—even with the best of intentions.

[16]"Trust Issues: Spotify's Commitment to Fans and Brands." Spotify for Brands. Accessed June 15, 2019. www.spotifyforbrands.com/en-US/insights/trust-issues-spotifys-commitment-to-fans-and-brands/.

[17]Saint, Nick. "Eric Schmidt: Google's Policy Is To 'Get Right Up To The Creepy Line And Not Cross It'." October 1, 2010. Last accessed February 16, 2020. www.businessinsider.com/eric-schmidt-googles-policy-is-to-get-right-up-to-the-creepy-line-and-not-cross-it-2010-10.

Bias in datasets

Ethical considerations and artificial intelligence date back to 1960 when Arthur Samuel wrote in *Science* about the moral consequences about a machine simply making conclusions from logical consequences of the input it is given.[18] Today, much of the focus on AI ethics is on the "what" (principles and codes) rather than on the "how" (practical application to AI). Ethics and AI have a long way to go.

> *Awareness of the potential issues [of AI] is increasing at a fast rate, but the AI community's ability to take action to mitigate the associated risks is still at its infancy.*
>
> —Morley, Floridi, Kinsey, and Elhalal (2019)[19]

HOW DOES AI KNOW WHAT IS IMPORTANT?

BOB: Let's take a medical example where AI takes in data and learns. One could argue that the very best data is from peer-reviewed journal articles. Studies described in these articles can be replicable (in theory), and medical science and careers advance through peer-reviewed publications.

GAVIN: But let's also consider generations of medical research from the 1960s and earlier where mostly men were participants. We have learned through the years that women have differing symptoms from men for the same disease. For example, women often delay seeking medical attention for a heart attack because they feel abdominal pain and not chest pain.[20]

BOB: This opens the question of whether the dataset used for AI applications adequately weighs evidence. The process of publications is to build on what is known. When a groundbreaking study is done, while it may be published in a top-tier journal,

---

[18]Samuel, Arthur L. (1960). "Some Moral and Technical Consequences of Automation—A Refutation." American Association for the Advancement of Science. 132(3429):741–742, 1960. https://doi.org/10.1126/science.132.3429.741.

[19]Morley, J., Floridi, L., Kinsey, L. & Elhalal, A. (2019). "From What to How: An Initial Review of Publicly Available AI Ethics Tools, Methods and Research to Translate Principles into Practices" *Science and Engineering Ethics*. December 11, 2019. Last accessed February 16, 2020. https://link.springer.com/article/10.1007/s11948-019-00165-5#Sec2.

[20]DeFilippis, Ersilia M. "Women can have heart attacks without chest pain. That leads to dangerous delays." Washington Post. February 16, 2020. Last accessed February 16, 2020. www.washingtonpost.com/health/women-can-have-heart-attacks-without-chest-pain-that-leads-to-dangerous-delays/2020/02/14/f061c85e-4db6-11ea-9b5c-eac5b16dafaa_story.html.

it takes time for more articles to be published to both replicate and further the science. And how does AI take groundbreaking results into its learning when a preponderance of articles exists on the older treatment?

GAVIN: Yep. When corrections to the science are made post AI learning, does the AI application get updated?

■ **The point**   How does the AI application "keep up with the literature" or simply stay current when new data come to light?

As an example, in 2018, the FDA fast-tracked and approved a new "tissue agnostic" cancer drug for a specific genetic mutation. Oncologists said that this new therapy would change the game, but how many studies need to be published until AI applications adopt it as the therapy of choice?

Researchers at the Memorial Sloan Kettering Cancer Center (MSKCC) who teamed up with IBM Watson sought to solve this question by creating "synthetic cases" that were put into training datasets so IBM Watson could learn from their data.[21]

BIAS FROM WHAT SOME BELIEVE TO BE TRUE

GAVIN: Essentially, MSKCC and IBM Watson added new cases to their dataset. They created records from their cases and placed them into research datasets containing other cases.

BOB: Presumably, this would make IBM Watson become smarter because it would have the benefit of MSKCC's knowledge. This is often referred to as the "Sloan Kettering way" of treating patients.

GAVIN: So, these "synthetic cases" were given to IBM Watson so it would learn. Doesn't this beg questions about whether these are common or unique cases, or even if MSKCC tends to receive a certain type of patient?

BOB: And because this was the "training set" where the AI modeled and *learned*, the bias can permeate future findings.

■ **The point**   Techniques to add "synthetic cases" to improve datasets may also add bias as well.

---

[21]Strickland, Eliza (2019). "How IBM Watson Overpromised and Underdelivered on AI Health Care." IEEE Spectrum. April 2, 2019. Last accessed November 6, 2019. https://spectrum.ieee.org/biomedical/diagnostics/how-ibm-watson-overpromised-and-underdelivered-on-ai-health-care.

We assume that peer-reviewed studies care for certain factors, such as representativeness, and control for bias or at minimum state them as assumptions/qualifications to the findings. When "synthetic cases" are created, one must ask questions such as these:

- Are these synthetic patient cases representative for the domain?

- Are they typical cases? Or are the edge cases?

- Did these patients transfer to the institution because they needed the worse/last resort treatment solutions?

- Is there potential for social, economic, racial, or gender bias in selection for these cases?

While this list merely nips around the edges of the potential for bias when an institution creates artificial or synthetic data to train AI, the need to apply ethical standards in AI becomes clear and apparent.

Let us be clear: MSKCC is one of the premiere cancer treatment centers in the world, but as Pilar Ossorio, a professor of law and bioethics at the University of Wisconsin Law School, argues, "*[AI] will learn race, gender, and class bias, basically baking those social stratifications in and making the biases even less apparent and even less easy for people to recognize.*" Considering that patients who are attracted to MSKCC tend to be more affluent, have a different mix of types of cancer, and have often failed multiple lines of treatment and are looking for one last chance,[22] these biases are woven into the very fabric of Watson's AI.

When the Watson team were pressed on concerns over the use of 'synthetic cases' to train IBM Watson, the response was striking. Deborah DiSanzo, general manager, IBM Watson Health, replied, "*The bias is taken out by the sheer amount of data we have.*[23]"

Considering how AI is a black box and we cannot truly know what data elements Watson's AI algorithm used or did not use, the relevance of the volume of data as an answer that overcomes potential bias is speculation at best.

---

[22]Gorski, D. (2019). "IBM's Watson versus cancer: Hype meets reality." *Science Based Medicine*. September 11, 2017. Last accessed February 16, 2020. https://science-basedmedicine.org/ibm-watson-versus-cancer-hype-meets-reality/.

[23]Strickland, Eliza (2019). "How IBM Watson Overpromised and Underdelivered on AI Health Care." IEEE Spectrum. April 2, 2019. Last accessed November 6, 2019. https://spectrum.ieee.org/biomedical/diagnostics/how-ibm-watson-overpromised-and-underdelivered-on-ai-health-care.

This is the issue with bias. It is often hard to see or incorporate into one's thinking. As an example, Dr. Andrew Seidman, who was the MSKCC lead trainer for IBM Watson, provided this answer to concerns over bias using MSKCC "synthetic cases" by proclaiming, "We *are not at all hesitant about inserting our bias, because I think our bias is based on the next best thing to prospective randomized trials, which is having a vast amount of experience. So it's a very unapologetic bias.*" This is why an ethical standard is needed and should be applied. It can be difficult for some to be objective.

TRAINING DATA SETS THE FOUNDATION FOR AI THINKING

GAVIN: The underlying concern is how pervasive bias can be when AI learns from a dataset with a questionable foundation. AI only learns what you feed into its training dataset. There is a lot more to successful AI than simply its programming.

BOB: Whether you bought the dataset and need to manage what is inside or you took the time to curate your own dataset, the data is critical to the process. Responsibility is on the product and data scientists teams to ensure good data hygiene.

GAVIN: Assume a result forms the basis for a product—one with AI at the core. How many corporations would retrain on a new dataset following the product launch?

BOB: There is a lot of risk on a complete retrain. What if the AI engine doesn't produce the same results with new training data? If it's bad enough, it could sink the product. There are a lot of companies or product teams that might not take that risk.

The point   Ethical standards are relevant today because AI is learning from datasets now. These datasets need to consider inherent bias or risk weaving that very bias into the foundation that is used to power the AI engine.

Toward an ethical standard

Organizations are concerned with the lack of an ethical standard for AI. In 2018, the MIT Media Lab at the Massachusetts Institute of Technology joined forces with the Institute of Electrical and Electronics Engineers (IEEE), a New Jersey-based global professional organization dedicated to advancing technology for humanity, and the IEEE Standards Association to form the global Council on Extended Intelligence (CXI). CXI's mission is to promote the responsible design and deployment of autonomous and intelligent technologies.

The IEEE welcomes engagement from those who wish to be part of the standards initiatives. The IEEE's Global Initiative's mission is, "To ensure every stakeholder involved in the design and development of autonomous and intelligent systems is educated, trained, and empowered to prioritize ethical considerations so that these technologies are advanced for the benefit of humanity."

This organization drafted a downloadable report entitled Ethically Aligned Design: A Vision for Prioritizing Human Well-being with Autonomous and Intelligent Systems, First Edition (EAD1e).[24] This report sets the foundation for an ethical standard for autonomous and intelligent systems. The IEEE P7000™ Standards Working Group standards projects listed as follows:

- **IEEE P7000** – Model Process for Addressing Ethical Concerns During System Design

- **IEEE P7001** – Transparency of Autonomous Systems

- **IEEE P7002** – Data Privacy Process

- **IEEE P7003** – Algorithmic Bias Considerations

- **IEEE P7004** – Standard on Child and Student Data Governance

- **IEEE P7005** – Standard on Employer Data Governance

- **IEEE P7006** – Standard on Personal Data AI Agent Working Group

- **IEEE P7007** – Ontological Standard for Ethically driven Robotics and Automation Systems

- **IEEE P7008** – Standard for Ethically Driven Nudging for Robotic, Intelligent and Autonomous Systems

- **IEEE P7009** – Standard for Fail-Safe Design of Autonomous and Semi-Autonomous Systems

- **IEEE P7010** – Wellbeing Metrics Standard for Ethical Artificial Intelligence and Autonomous Systems

- **IEEE P7011** – Standard for the Process of Identifying and Rating the Trustworthiness of News Sources

- **IEEE P7012** – Standard for Machine Readable Personal Privacy Terms

---

The point   There is an effort underway to develop ethical standards for AI.

---

[24]https://ethicsinaction.ieee.org/#set-the-standard.

Conclusion: Where to next?

So we have covered a couple of the concerns about the inputs to AI-enabled products. But we think there's another place where there is opportunity and what could keep AI applications from getting a bad rap: the user experience. One underlying theme that we touch on here and there is that there is a giddiness, an infatuation at times, with the technology that we forget that at the beginning and the end there is a user, a person. And, because we believe an AI application is just another application, we need to ensure that the AI application is tuned not only to the data but to the user's needs. The final chapter then describes the elements we feel will promote user engagement, and ultimately increase the likelihood of marketplace success.

Applying a UX Framework

A pathway for AI's success

When we look at the ground we have covered so far in this book, we see how AI and UX have some common DNA. Both started with the advent of computers and both with a desire to create a better world. We saw how UX evolved from a need to bring the information age closer to everyone.[1] AI grew similarly—with some fits and starts—and is now in the mainstream of conversation.

The benefits AI can bring are legion. However, there is a risk of another AI winter for reasons that have less to do with potential and more to do with perception. For many people, there's still a hesitance, a resistance, to adopt AI. Perhaps it is because of the influence of sci-fi movies that have planted images of Skynet and the Terminator in our minds, or simply fear of those things that we don't understand. AI has an image problem. Risks remain that people will get disillusioned with AI again.

---

[1]That task has been largely accomplished with the fact that more than five billion people have mobile phones, and over the half of those are smart phones. Smartphone ownership is growing rapidly around the world, but not always equally. Laura Silver. www.pewresearch.org/global/2019/02/05/smartphone-ownership-is-growing-rapidly-around-the-world-but-not-always-equally/ (retrieved April 16, 2020).

© Gavin Lew, Robert M. Schumacher 2020
G. Lew and R. M. Schumacher, *AI and UX*,
https://doi.org/10.1007/978-1-4842-5775-3_5

We believe that AI is ready. AI is more accessible than ever before and not just to the big players in industry, but within reach of many from startups to avid technophiles who are able to experiment with AI tools. This means AI is being embedded into new product ideas across almost all industries.

But, AI needs to be more than technology—the prescription for success is to not just embed AI into the product but also to deliver a solid user experience. We believe that this is the key to success.

NO ONE SETS OUT TO BUILD TERRIBLE EXPERIENCES

BOB: Let's get something straight. No company wants to build a product with a terrible experience.

GAVIN: But, think about it. One does not have to work too hard to remember experiences where you caught yourself saying, "What were they thinking when they made this?"

BOB: Several years ago, I used to give a talk that was more on the foundations of UX. Every application's user interface presents an experience. What we need to realize is that there is a designer behind the application and it's the designer's choice as to what that experience is: it can be amazing or a dud.

GAVIN: I don't really think that programmers wake up in the morning and say, "I'm gonna make things just a little harder for those users." But here's the thing, if programmers don't set out to create bad user experiences, why are there so many bad user experiences?

BOB: Yup, that is a paradox. There are a lot of reasons why so many products have such mediocre experiences—cost, awareness of users, time, laziness, and so on.

GAVIN: I would say though that most product owners underestimate how hard it is to design good experiences.

BOB: Luckily, technology has advanced to where microwaves and clock radios no longer default to flashing "12:00." Fixing the time after a power outage was so annoying. But, just this last Christmas, I spent hours trying to set up new gadgets for my house only to scream out in frustration.

GAVIN: I believe experiences matter. A product can promote the most amazing features, but thinking of AI-UX principles, when I set it up or use the product, is the interaction intuitive? Do I trust that the product works? Think of products you love and use every day—how much of that is because the user experience isfrictionless and enjoyable?

BOB: Technology has become a commodity. What can set a product apart is good design. The same logic applies to AI-enabled products.

The point Designing for simplicity is hard; it takes commitment.

What makes a good experience?

Everyone has these digital things scattered around the house in closets, basements, and workspaces. Think of a product recently purchased:

- Was it easy to order?
- Was it easy to set up?
- If you used instructions, did they help? (Were they even necessary?)
- Could you get the product to work quickly?
- Did it work like you thought it would?
- Do you still use it or does it sit idle after a month?

What is it about these products that fail? While there can be many reasons for why products do not meet expectations, all too often, we find ourselves shaking our head and asking,

What were they [designers, engineers, and product people] thinking when they designed this? It doesn't work the way I think.

Again, manufacturers do not set out to make products that disappoint—but they exist. Why? Sometimes, the simple answer is that the product creators did not spend enough time on the true need. Put in a different way, they built the product because they felt the technology was so compelling that they assumed everyone else would be captivated by it.

Often users find novel ways to use products far different from what the company or organization expected.

DO PEOPLE STILL BUY ALARM CLOCKS?

GAVIN: I believe watches have always been fashionable, but now, we don't use watches to tell time any more. We use them as fashion accessories. The functionality of my watch has been replaced by my mobile phone.

BOB: And your phone tends to be more accurate than your old Timex watch too!

GAVIN: This is a powerful example because the ubiquity of mobile phones really changed behavior. Think about early mobile phones. Some phones let you set an alarm as a feature. Now, you can set multiple alarms on your phone. I can even say, "Hey, Google, set an alarm at 7 a.m." And she replies, "Got it. Alarm set for 7 a.m. tomorrow."

BOB: More to the point, did the early phone manufacturers think that their product would have decreased sales for physical alarm clocks or result in people not wearing watches as often as they did 20 years ago?

The point   Even the most seemingly mundane features can change behaviors and change markets for products. When people use a product, their expectations and behaviors change in unanticipated manners.

Understanding the user

How do we better understand how people use products? A better understanding of the user experience is the answer. The ultimate purpose of a user-centered design (UCD) is to build products and services around users and their needs.

We have been involved in research and design of all manner of applications and products for decades. We have also seen many different phases and approaches to user interface design. The method that we see with most success is a user-centered design.

Definition   **User-centered design** (UCD) places user needs at the core. At each stage of the design process, design teams focus on the user and the user's needs. This involves a variety of research techniques to understand the user and is used to inform product design.

UCD[2] advocates putting the user and the user's needs as primary during each phase of the design process. While this seems obvious that one would design with the user in mind, the reality is that surprisingly few development efforts strictly follow this process end to end.

Research matters

It is hard to disagree with the idea that any product is better if the intended user has a good experience. Designing these seamless experiences is the result of hard work that maps what the user expects and needs into the design and interaction models for the product.

User research is the key method to capturing the needs of the intended user and integrating these insights into the product design.

---

[2]Another term coined by Norman in the title of his book: *User Centered System Design: New Perspectives on Human-computer Interaction* (1986). United Kingdom: Taylor & Francis.

HARD TO CALL A BABY UGLY

BOB: In order for UCD to really work, user research is needed. I have heard many creative and design directors say that "they *know* what the user needs!" But, the reality is that having evidence describing user needs is better than one's belief, however adamant one says it.

GAVIN: And at the very least, designers should be humble as they recognize that iteration can only make things better. Test early versions with prospective users. Get feedback early and often. Don't be afraid to be wrong. Know that the design will be better with this feedback.

BOB: And let someone else do the research. Let an objective party evaluate the initial designs. It is too easy to fall in love with something that you have poured time and energy into building. The product becomes "your baby," and let's face it—it is hard to hear someone call your baby ugly. You might ignore the criticism, or you might get defensive to protect it.

GAVIN: But this is the best thing to hear—especially early in the design process. The parts that are "ugly" appear as confusion or outright frustration when naive users interact with your product. Learn from what users think and improve the design.

The point   Make mistakes faster by having users interact with early-stage designs. Then, improve and repeat.

Does research really matter?

Sometimes, the best argument against a belief is a rhetorical question. The merits of user research are often put down by those who believe that creativity and innovative thinking does not need evidence-based UCD by repeating an adage often attributed to Henry Ford:

> If I had asked people what they wanted, they would have said faster horses.

The question "does research really matter?" places people into two groups: those who believe true innovation comes from gifted visionaries and those who believe that understanding what people think matters to sound design.

As a researcher, this is not even a good question because it is leading (i.e., directs the answer in a particular direction and is biased). While there are some who might be truly visionary, the reality is that all too often, research is the evidence that drives the innovation, the need that can be filled. Identifying this need and designing around it is where user research is best suited.

WHERE WERE THE HORSES?

BOB: When I think of the Henry Ford's often recited adage, I shake my head. Rhetorically, it implies insight cannot be obtained from the user. If Henry Ford had asked the question, he would have not made the Model T.

GAVIN: I think the time Henry Ford would have said this is in the mid-1920s. Think of *The Great Gatsby,* which was set at around the same time.

BOB: Whether in the book or in its theatrical interpretations, automobiles were prominent.

GAVIN: Exactly. Where were the horses? In *The Great Gatsby*, no one talked about horses. Driving and owning an automobile was a key theme.

BOB: Remember, Henry Ford did not invent the automobile. He invented a conveyor belt system to make cars more efficiently.

GAVIN: And that is why Henry Ford *never said that quote.*[3] What he did say was, "Any color, so long as it is black."

The point   While great innovations can indeed come from gifted designers, obtaining user feedback on designs early and often is a recipe for success.

Objectivity in research and design

User research is the key method to capturing the needs of the intended user and integrating these insights into the product design.

The UX lens

Many readers of this book may not have the skill (or inclination!) to debug Python code, but we believe that there is much more to AI than the code. Sufficient attention is usually given to the AI engine proper. AI-enabled products can benefit by shifting attention to everything *around* the AI engine. How can we improve everything else?

---

[3]Vlaskovits, P. (2011). "Henry Ford, Innovation, and That 'Faster Horse' Quote." Harvard Business Review. August 29, 2011. Accessed May 18, 2020. https://hbr.org/2011/08/henry-ford-never-said-the-fast.

LET'S FOCUS AWAY FROM THE AI OUTPUT AND ON THE EXPERIENCE

BOB: A common output of AI is a number or coefficient, such as 0.86: a correlation between 0 and1. Let's take your credit card fraud example. You're out to dinner and make a payment using your credit card...

GAVIN: So, as my payment transaction is processing, there is an AI program analyzing and thinking about fraud. In this example, AI outputs 0.86 and this means the transaction could be fraudulent (i.e., the algorithm assumes anything above .8 is likely to be fraud).

BOB: That is the extent of AI. The outcome was 0.86. But the *experience* with the AI-enabled fraud detection product is an alert to my phone in the form of a text message.

GAVIN: It could read, "WARNING. POTENTIAL FRAUD DETECTED ABOVE 0.80. CODE F00BE1DB."

BOB: Or someone could spend time designing a better interaction. The message could be much more friendly and provide the user with a call to action, such as "To authorize this purchase, Reply Yes."

The point   Work on what touches the user, such as messaging and user interactions. What AI provides can be quite amazing, but what makes the product a success is the experience.

We think AI can be seen through the lens of how we look at the user experience of any product or application. AI is no different. To be successful, it must have the essential elements of utility, usability, and aesthetics.

So what are the principles and what is that process?[4]

Key elements of UX

UX is not one thing for all users. It is multivariate. The various constituent parts, which we introduce next, combine to provide the user with an experience that is beneficial and maybe even delightful.

---

[4]We are not going to go into the details here. There are many websites and books on how to implement the UCD process into many different development approaches. A good place to start is here: www.usability.gov/what-and-why/user-centered-design. html (retrieved May 13, 2020).

Utility/functionality

Probably the most important thing that defines any application is what it does—we call this "utility" or "functionality" or "usefulness." Basically, is there a perceived functional benefit? In more formal terms, does the application (tool) fit for the purpose it was designed for? Does it do what the user needs? A hammer is good for pounding nails, not so good for putting on makeup. Any application needs to have the features and functions that a user expects and needs to be useful. These are table stakes for a successful product.

STRIVE TO CREATE FRICTIONLESS EXPERIENCES

BOB: Let's go back to the car example with a natural language voice assistant. The table stakes here is that AI needs to reliably understand human speech. In a car, it could be touchless controls like "call mom," "turn on the heater," "find a destination on the navigation system," and so on.

GAVIN: Earlier speech recognition in cars was very command driven and typically the user needed to know the exact voice command to say. Do you say, "Place call…" or "Make call…" or "Dial…" or "I want to call…" or even "Can you please call…"

BOB: The concept of using your voice has a benefit because the driver can keep their eyes on the road. But, too often, the rigid structure required the driver to remember the phrase. And, as the driver guesses commands, random noise can interfere. The driver might have been correct on the command, but the external noise caused the system to reply, "Sorry, I did not understand."

GAVIN: Instead, a better user experience would be to design an interaction that accepts many alternative commands in the presence of typical ambient noises found in cars.

The point Designing frictionless or effortless interactions can allow users to experience the utility and functionality. Early attempts to put AI-based speech recognition in cars had poor usage not because voice activation would not be beneficial, but possibly because the driver found it difficult to engage.

Usability

Utility and usability are often conflated, but they are independent constructs. Let's consider an absurd example, steering a car. We all can use a steering wheel to turn a car to the right or to the left. But if you think about it, there are other ways in which a car could be controlled. A keyboard where you would type "turn right" or a joystick or a remote control. There is a difference between function and how that function is implemented. Some of those ways are more usable than others.

Definition Usability is defined as whether a product enables the user to perform the functions for which it is designed efficiently, effectively, and safely.

Because car manufacturers aligned to a steering wheel, people who know how to drive can rent a car and, after some seat and mirror adjustments, drive a make and model of a car they have never driven before with a high degree of proficiency. These are standards that help guide how people interact with different instantiations of a system.

But what about new or systems requiring new controls? How do you create an **interaction** that is **usable**?

The basics of a **UCD process**, involve:

1. Early-stage research (exploratory or discovery) to capture user expectations and understand what people think.

2. Construction of a prototype.

3. Formative research, such as a **usability test**, to identify areas of confusion and gaps preventing a satisfactory experience.

4. Iterative design and more **usability tests** to further refine the product.

Definition A **usability test** is a qualitative study where intended users who are naive to the product are given context and asked to complete tasks for which the product was designed. Behaviors and reactions are observed. A usability test typically involves small sample sizes and feeds into an iterative design process.

More detail will be provided in the **Principles of UCD** section later in this chapter.

NATURAL GESTURES ARE A MYTH

BOB: Apple said swiping and pinching on a phone are *natural gestures* where people just *intuitively* know how to do it.

GAVIN: Really? I remember 2007 playing with my first iPhone. I saw the commercials by AT&T. This is where I learned to swipe and pinch.

BOB: Those commercials might have been the most expensive user manuals ever made!

GAVIN: Apple has patented scores of gestures that it calls natural. You have to take a look at Apple's patents like "multi-touch gesture dictionary."[5] My favorites are Apple's patents for *print* and *save*. For *print*, you put three fingers together, say, at the center, and branch out to form the points of a triangle. But to *save*, you do the opposite. Three fingers stay apart like points on a triangle and move inwards to the center. How is this something people know?

BOB: If these gestures were so natural, then why did Apple go to the trouble of patenting so many?

GAVIN: This is how myths are made. This is a 2009 study that describes the gestures that people know.[6] But, we did studies on multi-touch mobile phones in 2006 and 2007 where I saw people struggle. In 2009, people may have learned and adapted, but to say gestures have always been known and are natural is hard for me to reconcile from what I saw in the lab on the LG Chocolate, Palm, and first iPhone.

The point Not everything is natural or springs to mind when someone interacts with technology. Research and design go hand in hand to make the experience usable.

There are many products that have the same functions but those functions are rendered in different ways; some ways being more usable than others. An obvious example is booking a seat on an airline. Many sites will sell the exact same seat, they all do it a little differently, and some do it better than others. The function or purpose is the same, but the usability is different.

UX, AI, and trust

For a moment, let's consider one of our **AI-UX principles**: **trust**. One of the reasons that AI often fails to deliver for the users is that the outputs are off point. That is, we can't trust what we see or hear. The two dimensions of UX (utility and usability) speak directly to trust. If we build an app that lacks sufficient utility or suffers from poor usability, users will fail to trust the app and abandon. The same will happen with an AI app.

We find Alexa to have a lot of utility in some specific areas—giving the time, playing the news, telling the weather, and so on. And for these tasks, Alexa is also very usable. Thus, in a narrow, but important way, Alexa serves the user

---

[5]Elias, J.G., Westerman, W.C., & Haggerty, M.M. (2007). "Multi-touch Gesture Dictionary." USPTO US 2007/0177803 A1. Last accessed: June 22, 2020 www.freepatentsonline.com/20070177803.pdf.

[6]Wroblewski, L. (2010). Design for Mobile: What Gestures do People Use? Referencing Dan Mauney's Design for Mobile conference. Last accessed: June 22, www.lukew.com/ff/entry.asp?1197.

well. Yet, there are tens of thousands of skills available. We have tried dozens of skills that we thought would be useful, but they were deleted quickly because they were not usable. One new skill Bob tried to get sports scores, launched hisRoomba by mistake; needless to say, he appreciated the clean floor, but it wasn't the Roomba app that I launched. It's aggravating at best.

Admittedly, designing for voice is hard and, often, the design fails in two ways on usability; the natural language processing is simply not robust enough to handle many cues, and the voice designers have not done a good job of designing the dialog. Many of these voice skills also fail on utility because they do not sufficiently anticipate the users and the use cases. Thus, Alexa is relegated not to do much more than set a timer or gather information for a shopping list.

MISTAKES ARE OKAY FOR PEOPLE, BUT NOT SO WITH AI

BOB: We both started out in telecommunications. Back when phones were wired and could survive a hurricane. Even when the power went out in the neighborhood, your phone usually still worked. Back when we were at Ameritech, we did an internal study on dialing. What we found was that people were about 98% accurate when dialing any single number.

GAVIN: That is pretty accurate, but when dialing seven or now ten digits to complete a call, an error can occur. Doing the math, it happens maybe once or twice every dozen or so times you manually enter a phone number.

BOB: Precisely. Even with what seems to be a highly practiced task, errors happen. Who can say that when dialing a number from a piece of paper that they have *never* accidentally made a mistake?

GAVIN: Imagine a **voice assistant** that is always listening for that wake word. Even with a high accuracy rate, mistakes happen.

BOB: When a person makes a mistake, we say, "Oh, I fat-fingered the number." But humans are less forgiving of machines. If machines make "fat-finger" errors, we are more apt to say, "This thing sucks!"

The point   Because people have such a low tolerance for mistakes made by machines, **trust** is very important to the success of any AI-enabled product.

Trust has a very powerful influence on perceived success. Consider the challenge posed by autonomous driving cars. In 2017, there were over 6.4 million vehicle crashes in the United States. That is a crash every five seconds by human drivers. According to Stanford University, 90% of motor vehicle accidents are due to human error.[7] What is the potential benefit to autonomous vehicles? In California, there are 55 companies with self-driving car permits to conduct trials. From 2014 to 2018, there have been a total of 54 accidents from these autonomous vehicles, and according to an Axios report, all but one were due to human driver error—not the AI.[8]

But the problem is that humans have **trust** issues with autonomous vehicles. According to a AAA study, 73% of respondents expressed lack of trust in the technology's safety.[9] However, the thesis of the user experience playing a pivotal role in user acceptance still holds. "Having the opportunity to interact with partially or fully automated vehicle technology will help remove some of the mystery for consumers and open the door for greater acceptance," said Greg Brannon, AAA's director of automotive engineering and industry relations. The AAA study further stressed that "experience seems to play a pivotal role in how drivers feel about automated vehicle technology, and that regular interaction with advanced driver assistance systems (ADAS) components like lane keeping assistance, adaptive cruise control, automatic emergency braking and self-parking, considered the building blocks for self-driving vehicles, significantly improve consumer comfort level."

■ **The point**   Experience and trust matter. Positive experiences with AI-enabled driving features can improve **trust** in more advanced AI technologies.

Weirdness

As AI-enabled products proliferate, more information is available to the system to be analyzed. As AI thinks about your data, how should AI engage *without making it weird?*

---

[7]Smith, B. (2013). "Human error as a cause of vehicle crashes." The Center for Internet and Society, Stanford University. Traffic Safety Facts Annual Report Tables. National Highway Traffic and Safety Administration. December 18, 2013. Last accessed: May 19, 2020 http://cyberlaw.stanford.edu/blog/2013/12/human-error-cause-vehicle-crashes.

[8]Kokalitcheva. K (2018). "People cause most California autonomous vehicle accidents." Axios. August 29, 2018. Last accessed: May 19, 2020 www.axios.com/california-people-cause-most-autonomous-vehicle-accidents-dc962265-c9bb-4b00-ae97-50427f6bc936.html.

[9]Mohn, T. (2019). "Most Americans Still Afraid To Ride In Self-Driving Cars" Forbes. March 28, 2019. Last accessed: May 19, 2020 www.forbes.com/sites/tanyamohn/2019/03/28/most-americans-still-afraid-to-ride-in-self-driving-cars/#5803114632da.

By weird, we mean situations that can make human to AI interactions uncomfortable, awkward, or unusual.

WHEN AI GETS WEIRD...

GAVIN: Imagine your daily commute in your car. There is a lot of data that can be collected about your driving habits, from what you are listening to on the radio to your speed to route taken—all time and date stamped. The car even knows if you are wearing your seatbelt.

BOB: And based on the key fob you are using, it probably can distinguish between different drivers. With all of this data—all collected passively in the background while you drive to work.

GAVIN: Now, if an AI system could recognize patterns in your behavior. Logically, AI could use this insight to be proactive and save you time and effort. For example, say, there is a major traffic jam on your normal route to work, it could inform the driver and even suggest an alternative route.

BOB: I would expect there is high value in this *suggestion* based on AI constantly thinking about me and my commute to work. This would be a *frictionless* experience for the user.

GAVIN: But, it could suggest a lot of things from my patterns. Perhaps I go to the gym every other day after work. It could inquire if I want directions to the gym. But, where is the line on what to suggest? Say, I go to places that I would rather not have AI *offer suggestions?*

AI making a *recommendation* could freak me out. There needs to be a "weirdness scale" to designate what is appropriate and helpful from what is downright creepy.

 The point   AI has the potential to offer suggestions based on personal patterns of behaviors and habits, but where is the line between what is acceptable and unacceptable? This is a good example of why AI and UX need to be linked.

A weirdness scale

The concept of creating a continuum of inappropriate to appropriate actions is not about precision, but to present a user-centered mindset to what actions AI should and should not do. This weirdness scale would ground the product team to reflect on the appropriateness of AI suggestions. Patterns formed based on user behaviors, both explicit (e.g., a driver entering an exact address for directions) and passive (e.g., the car enters a shopping area and parks for 45 minutes before heading home), are target-rich opportunities for an

AI system. Knowing whether or not to trigger an action can be shaped and guided through a **UX lens**.

Definition   When a pattern is identified and your AI product is ready to offer a *recommendation,* the event can be called a **trigger**. The actions that follow can be shaped with the user's needs at the center.

Having a concept such as this **weirdness scale** during the design stage of any AI-enabled product has value. It can set the perspective for the team that there are **triggers** that should and should not be acted upon. It establishes for the product team that there are boundaries that should be considered. This can lead to **interactions** with the user to help refine the AI engine by capturing the times when the user explicitly pressed or said "No" to the recommendation. The team can brainstorm and anticipate the possible **triggers** to illustrate the breadth of the scale and to define "guardrails" that should be recognized.

When you consider the concerns over "Big Brother" fears that AI systems are always *watching* or *listening* (even when there is little evidence to support that **voice assistants** are actually "always listening"[10]), care must be given by the team designing the AI-enabled product to improve **trust** and subsequent adoption and usage. This is not about what AI can predict, but the responsibility of the team to the design the product using an AI-UX perspective.

The point   AI has the potential to offer suggestions based on your habits, but where is the line between what is acceptable and unacceptable? This is a good example of why AI and UX need to be linked and is in the hands of the team making the product.

Aesthetics/emotion

As humans, we prefer to use things that are aesthetically pleasing.[11] Imagine two websites that have the same features and same controls (i.e., same utility and same usability), but one has a better rendering than the other. Most people would prefer the one that looks nicer. There is some evidence to

---

[10]Levy, Nat. (2020). "Three Apple workers hurt walking into glass walls in the first month at $5bn HQ." Geekwire. February 24, 2020. Accessed May 20, 2020. www.geekwire. com/2020/alexa-always-listening-new-study-examines-accidental-triggers-digital-assistants/.

[11]In fact, Don Norman wrote a book on the subject. Norman, D. (2004). *Emotional Design: Why We Love (or Hate) Everyday Things.* Basic Books.

show that under these exact conditions, users will judge the better-looking site as being more usable, even though it has exactly the same functions and controls (the so-called **aesthetic usability effect**). The emotion from using a product has a profound effect on our perception of it.

Definition   The **aesthetic usability effect** describes how users attribute a more usable design to products that are more visually pleasing to the eye.

Clearly, designing to a high standard for the look and feel of a product or application is important. The problem comes when companies think that all they need to do is provide a beautiful product and ignore the utility and usability dimensions. The glitz may sell, but users will remember and punish that product in the market, especially when it comes time to making buying decisions again.

PUTTING LIPSTICK ON A PIG

GAVIN: When I go to the Consumer Electronics Show where tens of thousands of products are showcased, I always find it curious how companies try to capture your attention with the shiny new tech.

BOB: What strikes me as fascinating is the push to anthropomorphize products. Is there really a reason for these AI applications to be rendered with faces? But the reason they are, I think, is the designer wanted to evoke emotions in the user. Interacting with a plastic object that has big eyes and a mouth that speaks might be less intimidating than an orb that speaks.

GAVIN: There is nothing wrong with trying to tap into the emotional factor with aesthetically pleasing, anthropomorphic talking heads.

BOB: But, recognize that there is a UX hierarchy at play. Getting the foundation sound is key. Consider the usability of the application. Does it do as intended? Is there a perceived usefulness? These are the things that have to be done right because how something looks can only take you so far.

GAVIN: The experience has to be good. Or you are just dressing up another mediocre product. Know that the **aesthetic usability effect** only takes a product just so far.

The point   Focus on getting the foundation right and then apply the visual treatment to make the product stand out and set apart from the rest.

UX and the relationship to branding

This is a good place to inject an issue that we think about a lot: branding and its relationship to UX.

There is this interconnection between brand perception and user experience. Sometimes "brand" covers a lot of sins in the user experience;[12] sometimes a poor user experience damages the brand's value. We've come to realize just how influential a brand is to experience. We've seen amazing products that have no brand value get discounted and mediocre products from respected brands get lauded. Maybe it does not need to be said, but the most useful and usable don't always win. If brand marketers are pushing something that the product does not deliver on, then the market will react. Trust is eroded when the promises exceed the reality.

As we are increasingly commercializing the AI in the product, branding issues surface. Just saying something has a "powerful AI engine" does not guarantee a sale. It needs more; it needs UX.

UX DELIVERS THE BRAND PROMISE

BOB: UX intersects with marketing whether we like it or not. And one of the things that we have learned from marketing and branding is that it's easy to sell the sizzle, but you have to deliver the steak too.

GAVIN: That's why getting the user experience right is so important. Brand marketing can tell us how easy something will be or how a product will change our lives, but unless we actually experience that, it's all just talk.

BOB: Yep, UX is a delivery on that brand promise. If the user doesn't experience what the brand marketer sold them, then that mismatch will undermine the credibility of the brand.

GAVIN: AI has the same problem. Lots of hype, but until it delivers in a meaningful way to change lives, it will be just the sizzle without the steak.

The point   Designing a product is more than just the product itself. The landscape is littered with failures. Build a better product experience—one that allows AI to show what it can do.

---

[12]Think of how powerful the Apple brand is. Now think about it. It has notoriously bad usability. Much has been written on this, for example, the rise and fall of iTunes, Apple's most hated app. www.theverge.com/2019/6/3/18650571/apple-itunes-rip-discontinued-macos-10-15-ipod-store-digital-music-wwdc-2019. Porter, Jon. Jun 3, 2019 (retrieved May 21, 2020).

Principles of UCD

The key principle of user-centered design is that it focuses the design and development of a product on the user and the user's needs. There are certainly business and technical aspects to engineering and launching a product. The UCD approach does not disregard or discount the needs of the business or the technical requirements. However, **UCD** process places emphasis squarely on the user because the tradeoffs made at the business and technical level for timelines, specifications, and budgets are assumed. The UCD focus ensures that the user is not cast aside easily.

So let's turn to the key components: users, environment, and tasks.

The users

The importance of understanding the user's goals, capabilities, knowledge, skills, behaviors, attitudes, and so on cannot be underestimated. There may also be multiple user groups as well. Descriptions of each group should be documented (sometimes these descriptions are expressed as "personas)." One trap we see is that designers and developers assume they know the user. Or they assume they are just like the user. Many designs fail because the description of the user was nonspecific, nonexistent, or just wrong.

The point   Define the intended user—beyond simply a market description or segment. Construct a persona describing the user's knowledge, goals, capabilities, limitations and include user scenarios that will both define the experience to be built and include "guardrails" of things to avoid.

The environment

Where are the places and what are the conditions where the users interact with the product? Environment includes the locations, times, ambient sounds, temperatures, and so on. For instance, certain environments (i.e., contexts) are better for some modalities (hands-free) than others. An app to be used on a treadmill while running will have different design characteristics than an app used for banking. What this means is that the user experience extends beyond simply what the user sees on the screen. The UX is the whole environment; the totality of the context of use. In registering for a new application that requires mobile phone use for two-factor authentication, for instance, assumes the user has access to the phone and must account for the possibility that the phone is not present or unable to be used. The whole process is part of the UX. The instruction guide that comes with a new smartphone is part of the UX. These things are not disconnected in the mind of the user, but they are often handled by different groups in the organization.

▨ **The point**   Develop use cases against the environments of use to identify further user needs. This could also illuminate the points where passive and explicit user data can be used to **trigger** additional user benefits.

IMPORTANCE OF ENVIRONMENT

GAVIN: In 2017, France's official state railway company, SNCF, wanted to build an AI chatbot ticket application.[13] The design team captured the conversations between travelers and ticket agents. They trained the AI's grammar system to model the experience observed.

BOB: But when they tested the chatbot prototype with users, it failed. When customers encountered the blank text field, they began with, "Hi. I would like to buy a ticket." Users expected to engage the chatbot in a friendly and conversational manner.

GAVIN: So, the chatbot expected, "I would like to buy a ticket from Paris to Lyon today at 10:00 a.m." Just like how they were observed at the ticket window.

BOB: Yep, they never encountered a friendly and conversational experience when they observed transactions at the ticket window!

GAVIN: Ah, this is Paris, right? There was probably a line of customers all waiting impatiently. If a customer walked up and said, "Hi. I would like to buy a ticket from Paris to Lyon or maybe Normandy... What times are available?" Those in the queue would be visibly angry. Such that the next person would walk right up and make the request as precise and efficient as possible!

BOB: Exactly! The environment of use was different. A chatbot could be used at one's leisure without the pressure from a queue of Parisian travelers!

▨ **The point**   The environment of use can change user interactions dramatically. Luckily, testing identified the issue and the team was able to retrain the chatbot.

---

[13]Lannoo, Pascal & Gaillard, Frederic (2017). "Explore the future: when business conversations meet chatbots." 13th UX Masterclass. April 22, 2017. Shanghai, People's Republic of China.

The tasks

Tasks are about what people do to accomplish their goals and involve breaking down the steps that users will take. Depositing a check using a mobile app requires me to log in, to navigate to the right place, enter amounts, take pictures, and a lot more. Tasks can be expressed as a task analysis or a journey map (if detailed enough). Based on a full explication of the tasks, the application requirements can be built.

The point   Use cases are well defined in product development processes. With AI, where the system presumably can learn and identify new opportunities, design for AI-enabled products now has AI-originated use cases which are not defined by the product team. Tasks have an added dimension with AI that needs to be managed.

While much of this can seem elementary, it amazes us how few organizations take the time to describe users, environments, and tasks. Getting to this detail requires user research[14] and must be documented in the specifications for the design and development team. The three key elements of UCD form the foundation of knowledge, but the process is what delivers a successful product or application that allows AI to show what it can do.

Processes of UCD

There are multiple flavors of this, but Figure 5-1 shows the basics of how the process works. Starting at the top and moving clockwise, early-phase user research would have defined the users, refined their needs, identified the context (i.e., environment) of use, and documented tasks in the requirements. At this point, we are ready to start design. Often the tendency among the technical team is to start programming. In UCD, we want to resist that urge. This is supported in *The Mythical Man-Month*[15] by Fred Brooks where he writes, "By the architecture of the system, I mean the complete and detailed specification of the user interface."

---

[14]Sometimes organizations will rely on market research for this kind of analysis. Market research and insights can be helpful, but they do not substitute for understanding of the user and user's needs at the level required for system specifications.

[15]Brooks, F. P., Brooks, F. P. (1975). The Mythical Man-Month: Essays on Software Engineering. United Kingdom: Addison-Wesley Publishing Company.

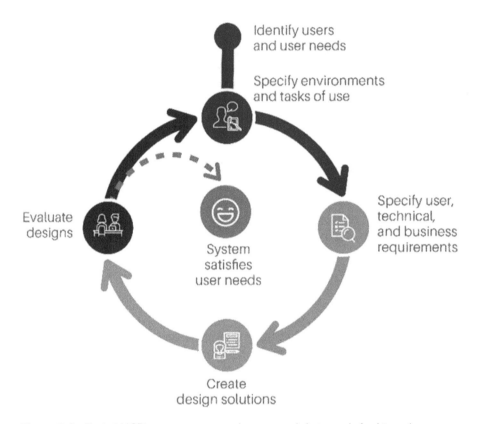

Figure 5-1. Typical UCD processes ensure the users and their needs feed into the process and then cycle through design and evaluation processes

Taking the research and building the user interface begins next.[16] Design should begin by defining with rough sketches so as not to get too wedded to early ideas.[17] Be willing to throw out a dozen ideas before settling on one. Draw the wireframes and map out the interactions—do they make sense? Do they get users closer to their goals?

Once design begins, conduct usability tests with paper prototypes. Successively refine them with representative users doing typical tasks. Later develop digital prototypes to test dynamic interactions, getting the design closer to the final

[16]We don't want to imply that business needs or technical needs are secondary; there is always a balance between user needs, business needs, and technical capabilities. We strongly believe that the business and technical teams should be a part of the UCD process.

[17]While we are introducing many of the concepts of UCD and user interface design for the purposes of showing how they can be beneficial to AI, we are not intending this to be a UCD book. There are hundreds of books, articles, and resources out there to get you started.

look and feel. There are countless books and websites describing methods on UI design, conducting usability testing and iterative prototyping that we're not going to cover here. Suffice it to say, usability testing is the single-most important component to perfecting the user interface.

As Figure 5-1 illustrates, this is an iterative process, a successive approximation to a system that meets the needs of the user. Once the design is evaluated, based on user feedback, one might adjust the requirements, decide on different ways to break down the tasks, or change the interface controls. A lot can change.

One key thing is that it is only at the end that we advocate applying graphical and visual treatments. What most people think of as design, graphical treatment, comes only after we have nailed the interaction models in the prototype designs. Colors, shape, size, and others, all support the underlying design. In construction, you can't paint the house before the walls go up. In application development, you should not create pretty diagrams first. Design begins in the field with research, not in Photoshop.

A best practice at the outset of design is to describe tangibly what success looks like as it relates to the user needs. For instance, you might set a goal that 95% of users get through the registration page in 2 minutes without error. These targets on critical tasks provide measurable progress that the development team can aim for. More importantly, success on these targets during usability testing is what breaks us out of that loop and moves us to the technical development.

The real benefit of the UCD model is that through progressive improvements documented in usability testing, the organization can have much higher confidence in the success of the final deliverable than if there was no, or only casual, feedback from users. Done well, applying the UCD model will make applications useful, usable, learnable, forgivable, and enjoyable.

UCD is agnostic to the content; it can, and should, be applied to AI applications.

A cheat sheet for AI-UX interactions

There has been some very good work exploring the relationship of the interaction of humans and AI from Microsoft and the University of Washington.[18] The research team there did a very thorough review of guidelines and vetted them experimentally. As you can see in Table 5-1, these guidelines are mostly geared toward improving the utility and usability of the AI interface.

---

[18]Saleema Amershi, Dan Weld, Mihaela Vorvoreanu, Adam Fourney, Besmira Nushi, Penny Collisson, Jina Suh, Shamsi Iqbal, Paul N Bennett, Kori Inkpen, et al. 2019. Guidelines for human-AI interaction. In Proceedings of the 2019 CHI Conference on Human Factors in Computing Systems. ACM, 3.

Table 5-1. Guidelines for AI-UX interactions from Microsoft and the University of Washington

#	Guideline	Description
1	Make clear what the system can do.	Help the user understand what the AI system is capable of doing.
2	Make clear how well the system can do what it can do.	Help the user understand how often the AI system may make mistakes.
3	Time services based on context.	Time when to act or interrupt based on the user's current task and environment.
4	Show contextually relevant information.	Display information relevant to the user's current task and environment.
5	Match relevant social norms.	Ensure the experience is delivered in a way that users would expect, given their social and cultural context.
6	Mitigate social biases.	Ensure the AI system's language and behaviors do not reinforce undesirable and unfair stereotypes and biases.
7	Support efficient invocation.	Make it easy to invoke or request the AI system's services when needed.
8	Support efficient dismissal.	Make it easy to dismiss or ignore undesired AI system services.
9	Support efficient correction.	Make it easy to edit, refine, or recover when the AI system is wrong.
10	Scope services when in doubt.	Engage in disambiguation or gracefully degrade the AI system's services when uncertain about a user's goals.
11	Make clear why the system did what it did.	Enable the user to access an explanation of why the AI system behaved as it did.
12	Remember recent interactions.	Maintain short-term memory and allow the user to make efficient references to that memory.
13	Learn from user behavior.	Personalize the user's experience by learning from their actions over time.
14	Update and adapt cautiously.	Limit disruptive changes when updating and adapting the AI system's behaviors.
15	Encourage granular feedback.	Enable the user to provide feedback indicating their preferences during regular interaction with the AI system.
16	Convey the consequences of user actions.	Immediately update or convey how user actions will impact future behaviors of the AI system.
17	Provide global controls	Allow the user to globally customize what the AI system monitors and how it behaves.
18	Notify users about changes.	Inform the user when the AI system adds or updates its capabilities

---

The point   Consider developing and applying a set of guidelines, such as those in Table 5-1, when designing your product's AI and user interactions. It covers a wide range of topic areas with UX implications.

A UX prescription for AI

We have looked at what some of the problems with AI have been in the past. We have also investigated why similar problems might manifest again in the future and why UX offers an effective solution to these problems. Now we will look at the "how": how can we use UX to avoid some of the pitfalls of AI?

The answer lies in advancing a framework for incorporating UX sensibilities into AI.

We are going to illustrate the goal by example. Let's say we want to build a smart chatbot into a desktop application (say, an electronic health record (EHR) used by doctors) to offer in-context help or clinical support. The first stop is to understand if this is a need that user's really have? If so, what do we know about those users?

Understanding users, environments, and tasks

Firstly, we'd want to understand our users; we'd conduct user interviews to understand how they use their EHR today. This will uncover the user's mental model for how they interact with the EHR. Workflows will be identified that will help set the stage for how AI would integrate seamlessly into daily tasks already performed. Moreover, the research will find existing difficulty and inefficiencies with the EHR which can be opportunities for AI to be more beneficial.

Example research areas of interest:

- What are the things you did in your job earlier this week? Ideally,
  - Shadow and observe actual activities; watch the tasks users do and capture the workflow with emphasis on the different sources of information and manual effort.
  - Watch how the user navigates and interacts with systems.
  - Pay attention to the types of help used.
- Who are the various users? Do they all need/want help for the same things? In the same ways?

To avoid being too influenced by a small set of users, in a case like this, we might use the qualitative interviews and observations to develop a survey to get more information on how users would be best supported by an EHR chatbot. The survey would collect data about the characteristics of the user, where help is needed in the EHR, under what circumstances help might be welcomed, and so on. Often we will take the quantitative data about users and identify groups with similar characteristics using multivariate statistical methods. These groups form the basis of personas, and we will carry out further, more detailed, interviews with people who fall into those groups to flesh out the personas. Our emphasis through all this is to get at the knowledge, skills, expertise, and behaviors of the users to know how best to develop the application to serve them.

Note that after doing these interviews, we may discover that the proposed construction of the chatbot is not a good solution and that users would be better aided in other ways.

Next, we move on to understanding the contexts of use, the environments. EHRs are used in exam rooms, in hospitals, at reception desk, in homes, and many more. Knowing what users do in each of these settings is important to know the kind of context-sensitive help that the chatbot would offer. If the chatbot offers the same kind of help to all users in all contexts, it's likely not to be so useful. Getting information about the environments might be able to be done in the same interviews as those for the users. Better still would be to include some questions in the preceding questionnaire to get a feeling for the environments, but then to go and observe the users *in the actual use environments*. Seeing how users use the application and all the supporting materials and procedures in *context* is essential to documenting requirements. This means going to the clinic and watching as many representative users use (and get frustrated with) the application.

Last, while gathering information about the users and their environments, data about the specific tasks people do can be gathered. If we know what each groups' goals are in using the EHR and learn about the environments, then we can begin to identify the various steps in the process to support them.

Underlying all this is the AI model—presumably based on thousands of cases—knows the interaction sequences (e.g., screens viewed), user inputs, error messages, and so on that occur. The AI model can infer where users might need the support of an intelligent bot. While these inferences are probably directionally correct, they can only be refined by the knowledge gathered from users. Perfecting these models may require tuning through crowdsourcing of the inputs and outputs. (More on this in the next section.)

Applying the UCD process

Armed with this background on users, environments, and tasks, initial design can begin. This design should start with the interaction model, controls, and objects, but at a rough level. Recall that the purpose is to help the user by monitoring the user's behavior and offer advice/support when it detects trouble. Design specifically when the chatbot should **trigger** and interrupt the user with important information or when decisions need to be made. Set the *guardrails* defining when and how that the chatbot interacts.

This is very tricky and potentially dangerous as distracting someone during a critical task on an application like an EHR might cause the user to forget to enter critical information.

The design team will go through a series of concepts, perhaps on paper or crude digital prototypes (e.g., using PowerPoint). It is here that the iterative design, test, revise, test cycle takes over. The design is put through usability testing with representative users doing representative tasks. Giving users the chance to experience how the chatbot interface will respond to actions will be essential in successively refining its behavior.

The important thing to note is that, up to this point, we are assuming that the algorithms are sound and that the underlying AI is appropriate. During user testing, not only should the experience of interacting with the chatbot be tested, but a test of the AI-based content should be conducted too. The point here being that once the design cycle begins, all aspects of the user experience should be in play. The user testing begins cheap and fast and often ends with a larger-scale more formal user test.

POOR CLIPPY—NOT!

GAVIN: Microsoft unleashed Clippy on the world in the mid-1990s trying to make using Office applications more friendly. It had a rules-based engine that would detect if you needed help and intervene.

BOB: Yeah, but it would totally stop my work and make me respond to it. It quickly became hated by most of us. It's even a punchline now. People have a nostalgia for it!

GAVIN: What's interesting though is that despite the annoying experience, it had another side. Microsoft ignored their own research—many women thought that Clippy was male, too male. And not only that, he was a creepy male.[19] The thing is, men did not feel this emotion—but women did.

---

[19]Clippy might never have existed if Microsoft had listened to women. Perkins, Chris. June 25, 2015. https://mashable.com/2015/06/25/clippy-male-design/. Retrieved May 15, 2020.

BOB: Clippy was an anthropomorphized agent that went wrong. It simply could never deliver the kind of support or build the kind of rapport with the user that was intended. He resurfaced through the years most recently in Microsoft Teams, but that was crushed in very short order![20]

GAVIN: Yeah, Clippy's become an exemplar of how not to annoy your users. So many poor decisions!

The point User research can identify potential trouble and allow for course corrections to solve issues.

In a development like this, the measures of success are perhaps a little harder to come by. Clearly, user measures of utility and usability are high on the list. We see this today in many applications—users get asked after many interactions how many stars they would give the quality of service. Other objective measures can also be taken. If this is supposed to be a user assistant, we could measure whether the number of successful actions after the chatbot interaction has increased or not. It is important to have these measures established so that we know we have succeeded in the design and can proceed with detailed programming.

What else can UX offer? Better datasets

We've spent a lot talking about the benefits of UX and the UCD process to AI. But there may be other ways UX can benefit AI.

User experience comes from various disciplines within psychology. One important thing that psychology and therefore user experience can bring to AI is a set of capabilities for collecting better data.

Giving AI better datasets

As we discussed earlier in Chapter 4, one of the biggest drawbacks for AI is getting the right data. Many in UX researchers are trained in research methods to collect and measure human performance data. UX researchers can assist AI researchers in collection, interpretation, and usage of human-behavior datasets for incorporation into AI algorithms.

What is the process by which this is done?

---

[20]Microsoft resurrects Clippy and then brutally kills him off again. Warren, Tom. March 22, 2019. www.theverge.com/2019/3/22/18276923/microsoft-clippy-microsoft-teams-stickers-removal. Retrieved May 15, 2020.

Identify the objective

The first task is to understand what the AI researchers really need. What is the objective? What constitutes a good sample case? How much variability across cases is acceptable? What are the core cases, and what are the edge cases? So, if we wanted to get 10,000 pictures of people smiling, is there an objective definition of a smile? Does a wry smile work? With teeth, without teeth? What age ranges of subjects? Gender? Ethnicity? Facial hair or clean shaven? Different hair styles? And so on. Both the in and out cases are components that the AI researchers need to clearly define and have all parties agree on.

Collect data

Next, do the necessary planning for data collection. One of the strengths of UX researchers is the ability to construct and execute large-scale research programs that involve human subjects. How to collect masses of behavioral data face to face, efficiently, and effectively is not in the core expertise of many AI researchers. In contrast, much of the practice of user research is about setting the conditions necessary to get unbiased data. Being able to recruit sample, obtain facilities, get informed consent, instruct participants, and collect, store, and transmit data is essential. Furthermore, UX researchers can also collect all the metadata necessary and attach that data to the examples for additional support. UX researchers are practiced in sorting, collecting, and categorizing data—as is evidenced by a skillset that includes qualitative coding and the many tools that support these types of analysis.

Do further tagging

After initial data collection, it may be necessary to organize and execute a crowdsourcing program such as Amazon's Mechanical Turk[21] to further augment the data collected so far. For instance, if we were to collect voice samples of how someone orders a decaf, skim, extra-hot, triple-shot latte in a noisy coffee shop, there could be several properties that would be of interest for each sample. In such cases, we might engage multiple researchers, or coders, to review each sample, transcribe the samples, and judge their clarity and completeness. These coders would then have to resolve any observed differences to ensure the cleanliness of the coding.

---

[21]"Crowdworkers" or "Turkers" do work that machines cannot do, yet. A review of the use cases of MTurk reveals that there is a wide range of work available to Turkers that supports machine learning datasets. There are potentially as many as 500,000 turkers worldwide. AI needs humans.

```
┌─────────────────────────────────────────────────────────┐
│          COLLECTING DATA IN REAL WORLD                   │
└─────────────────────────────────────────────────────────┘
```

BOB: We recently did a study where we had to get representative users to enter voice commands into a mobile device in a way they would normally speak to a colleague.

GAVIN: Yes, we sampled more than a thousand people through an on-line survey, gave them scenarios, and they had to respond with the commands. They spoke the commands into their mobile phone. We were dealing with all sorts of accents, so the voice recognition was a challenge.

BOB: Yeah, and there was lots of ambient noise. The speech then had to be converted to text. There was a lot of really good content and voice capture, but there was also gibberish. But, that's realistic.

GAVIN: Ultimately, there were dozens of coders who listened to the recordings and checked the speech to text transcriptions in order to tune the AI algorithms. It was hard work, but necessary to make sure that the AI engines get it right in the future.

The point   AI-enabled apps often need a lot of humans to curate the input.

Well-constructed and executed research programs can help protect against the limitations that may be present in existing databases used to train AI algorithms. Using custom datasets avoids the use of inconclusive, useless, or incorrect data whose limitations might not be immediately obvious. UX researchers are well positioned to help ML scientists collect clean datasets for the training and testing of AI algorithms.

Similarly, one increasing objection to AI is that it has been trained on biased data; we touched on this earlier. By controlling the sample from which the data are collected, datasets can avoid that bias. We once did a large-scale study ($n=5000$) for a company that wanted to collect videos of people doing day-to-day activities. One of the key criteria for them was to get a representative sample of the population—by age, gender, ethnicity, and so on—so that their facial recognition algorithms could be trained on better data and their output more accurate.

Where does this leave us?

Find the why

We opened the book with a "Hobbit's tale" suggesting AI and UX have been on quite a journey. One filled with hype and periods of winter. The future for AI is quite immense. In the not too distant future, we will witness

AI integrating into almost every industry to bring forth better health, liberation from mundane or dangerous jobs, and advances beyond what can be imagined. However, we believe that success will come more readily and may avoid failures if more attention were placed on the user experience to specifically improve AI-UX principles of context, interaction, and trust. We believe AI-enabled products need to not be singularly focused around technology. As a solution, we recommend using a UCD process—one that places the human, who is the beneficiary of AI's progress, at the core. However, we would like to end with one additional suggestion—to build AI-enabled products with a purpose, find the *why* that will give AI purpose that will drive the design to greater success.

FINDING THE WHY

GAVIN: You know, I started my research career at UCSF studying brainwaves. I placed electrodes on people's heads and recorded electrical impulses from the brain when the participant heard a "boop" or a "beep" or when they lifted their index finger.

BOB: You were a lab technician doing basic research!

GAVIN: Exactly. The year was 1991, and during one research session, the participant finished, and as he left, he shook my hand and said, "I really hoped this research goes to find a *cure*." I smiled and said, "I hope so as well." When the participant left, I was devastated.

The participant was HIV+, and at the time, retrovirus treatments were still experimental. I thought that in 10 years, this 21-year-old, bright-eyed, and energetic person might have full-blown AIDS and probably would pass away before a cure is found.

BOB: You were doing basic research. Comparing his brainwave activity to see if it resembled an Alzheimer's patient or alcoholic's when a "beep" is heard…

GAVIN: Or when an index finger is raised…

BOB: We still haven't found a cure for any of those three diseases and that was almost 30 years ago.

GAVIN: You could say that was the day I lost my purpose. I moved into UX research where I hoped research would have more direct impact for me.

Fast forward 10 years. I was with a patient doing a study on a prototype for an auto injection device. I remember the patient needed to pause and stand up because she had such terribly fused joints from her disease. At the end of the session, she did not shake my hand like the HIV patient. She gave me a hug.

BOB: I have never received a hug in a research session before!

GAVIN: Me either! I was surprised. She looked at me and said, "You don't get it, do you?" I shook my head. She continued, "Look at my hands and see how hard it is for me to hold things. Currently, I have to reconstitute this medicine. The process is so complicated that I have to do it all on my kitchen table. I have to syringe one drug and mix with another precise amount. Wait and then inject. Again, my hands barely work, but this therapy stops the progression and it is a miracle drug for me."

I said, "The purpose is to understand how it works for you. We want to understand how we can make the experience match your expectations so you can have a safe and effective dose."

She replied, "So I can use this new device for the first time correctly? I can walk into the bathroom and be done in a minute?"

I said, "Exactly."

She shook her head and said, "You still don't get it. You see, my current process is so complex that I do it on the kitchen table and it takes forever. Well, my 5-year-old daughter sits and watches her mother struggle to take her miracle medicine. Now, with this device, I can take my medicine discreetly in a bathroom.

"Well, you are not just making this safe and effective for people like me," she said. "You are changing the way a daughter looks at her mother! My daughter does not have to watch her mom struggle with her medicine and shoot up to live."

BOB: Wow. When a product is done well and is appreciated, the reasons can be broader than imagined.

The device's *purpose* for this patient is powerful.

GAVIN. I found *my purpose*. Companies need to also find purpose within a product and let this help drive better design.

The point   Have a strong understanding of how the user experience will be affected by its usage. Who will be affected? Identify situations of use that can amplify the user benefit. A product's *purpose* will be found there. This is the "why" that will drive the product team to design a compelling experience rather than falling prey to the hype surrounding the technology.

Our hope is that a greater recognition of the symbiotic relationship between people and AI can be designed into products. We want AI to become less of a black box and more transparent about its strengths and weaknesses. UX can help. AI development efforts should involve the user and user's goals directly, be aware of the environments, and account for the tasks. We believe that AI products with a user-centered focus will be hands down more successful than those that do not have that focus. Remember,

If AI doesn't work for people, it doesn't work.

Index

I

A

© Gavin Lew, Robert M. Schumacher 2020
G. Lew and R. M. Schumacher, *AI and UX*,
https://doi.org/10.1007/978-1-4842-5775-3

CPSIA information can be obtained
at www.ICGtesting.com
Printed in the USA
LVHW021517281221
707330LV00004B/142